6th Edition

CRISC

Review Manual

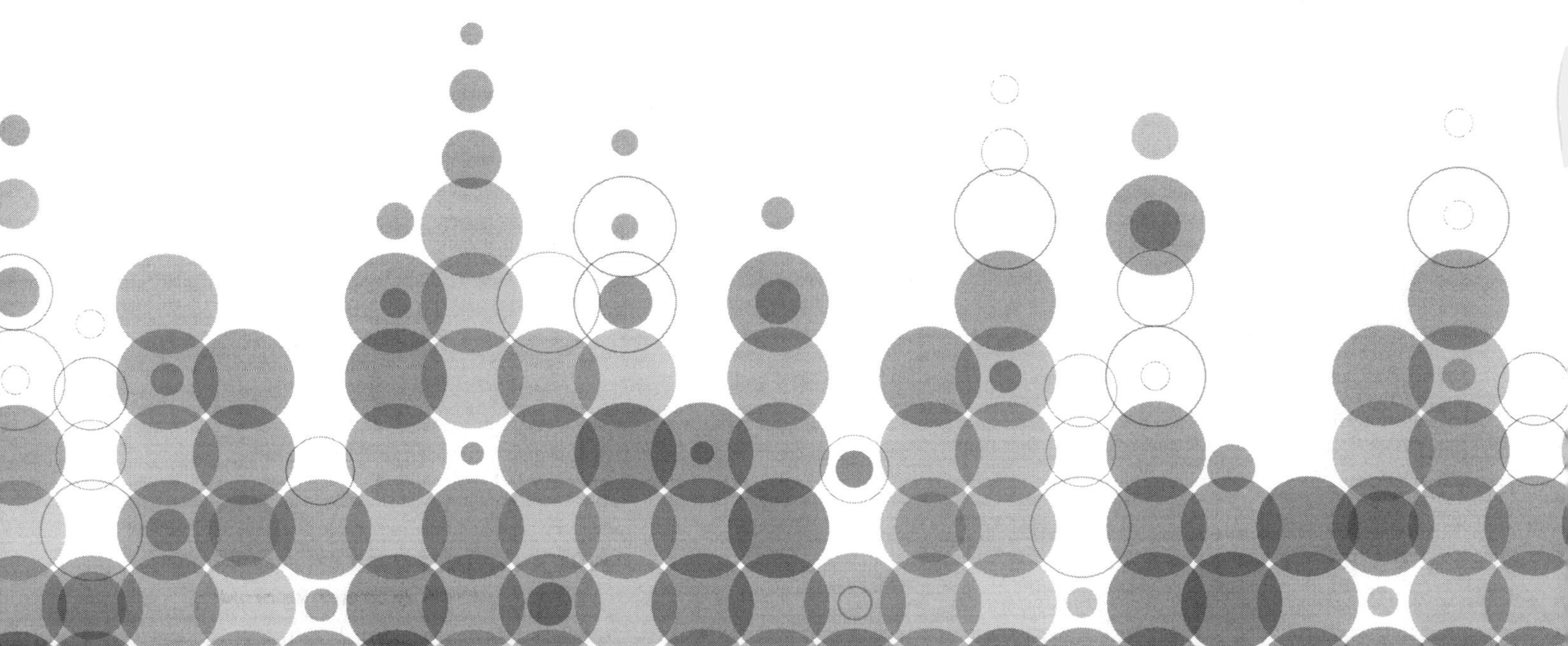

ISACA®

ISACA (isaca.org) helps global professionals lead, adapt and assure trust in an evolving digital world by offering innovative and world-class knowledge, standards, networking, credentialing and career development. Established in 1969, ISACA is a global nonprofit association of 140,000 professionals in 180 countries. ISACA also offers the Cybersecurity Nexus™ (CSX), a holistic cybersecurity resource, and COBIT®, a business framework to govern enterprise technology.

In addition, ISACA advances and validates business-critical skills and knowledge through the globally respected Certified Information Systems Auditor® (CISA®), Certified Information Security Manager® (CISM®), Certified in the Governance of Enterprise IT® (CGEIT®) and Certified in Risk and Information Systems Control™ (CRISC™) credentials.

Disclaimer

ISACA has designed and created *CRISC™ Review Manual 6th Edition* primarily as an educational resource to assist individuals preparing to take the CRISC certification exam. It was produced independently from the CRISC exam and the CRISC Certification Working Group, which has had no responsibility for its content. Copies of past exams are not released to the public and were not made available to ISACA for preparation of this publication. ISACA makes no representations or warranties whatsoever with regard to these or other ISACA publications assuring candidates' passage of the CRISC exam.

ISACA

3701 Algonquin Road, Suite 1010
Rolling Meadows, IL 60008 USA
Phone: +1.847.253.1445
Fax: +1.847.253.1443
Email: *info@isaca.org*
Web site: *www.isaca.org*

Participate in the ISACA Knowledge Center: *www.isaca.org/knowledge-center*
Follow ISACA on Twitter: *https://twitter.com/ISACANews*
Join ISACA on LinkedIn: ISACA (Official), *http://linkd.in/ISACAOfficial*
Like ISACA on Facebook: *www.facebook.com/ISACAHQ*

ISBN 978-1-60420-371-4
CRISC™ Review Manual 6th Edition
Printed in the United States of America

CRISC REVIEW MANUAL 6TH EDITION

ISACA is pleased to offer the sixth edition of the *CRISC™ Review Manual*. The purpose of the manual is to provide CRISC candidates with information and references to assist in the preparation and study for the Certified in Risk and Information Systems Control™ (CRISC) exam.

The *CRISC™ Review Manual* is updated regularly to keep pace with rapid changes in the identification, assessment, response, monitoring and reporting of risk and information systems (IS) controls. As with previous manuals, the sixth edition is the result of contributions from many qualified authorities who have generously volunteered their time and expertise. We respect and appreciate their contributions and hope their efforts provide extensive educational value to CRISC manual readers.

Your comments and suggestions regarding this manual are welcome. After taking the exam, please take a moment to complete the online questionnaire (*www.isaca.org/studyaidsevalutation*). Your observations will be invaluable for the preparations of the seventh edition of the *CRISC™ Review Manual*.

The sample questions contained in this manual are designed to depict the type of questions typically found on the CRISC exam and to provide further clarity to the content presented in this manual. The CRISC exam is a practice-based exam. Simply reading the reference material in this manual will not properly prepare candidates for the exam. The self-assessment questions are included for guidance only. Scoring results do not indicate future individual exam success.

Certification has resulted in a positive impact on many careers, including worldwide recognition for professional experience and enhanced knowledge and skills. The Certified in Risk and Information Systems Control™ certification (CRISC™, pronounced "see-risk") is designed for IT and business professionals who have hands-on experience with risk identification, risk assessment, risk response, and risk and IS control monitoring and reporting. ISACA wishes you success with the CRISC exam.

ACKNOWLEDGMENTS

The *CRISC™ Review Manual 6th Edition* is the result of collective efforts of many volunteers. ISACA members from throughout the globe participated, generously offering their talent and expertise. This international team exhibited a spirit and selflessness that has become the hallmark of contributors to ISACA manuals. Their participation and insight are truly appreciated.

Special thanks go to James Samans, CISA, CISM, CRISC, CISSP, CIPT, PMP, XENSHA LLC, USA, who worked on the sixth edition of the *CRISC™ Review Manual*.

Expert Reviewers
Alvaro Rodrigo Cayul, CISM, CRISC, Chile
Robert T. Hanson, CISA, CISM, CRISC, Australia
Ken Hendrie, CISA, CISM, CGEIT, CRISC, GCIH, ITIL, PRINCE2, HackLabs, Australia
Shawna M. Flanders, CISA, CISM, CRISC, Business - Technology Guidance Associates, LLC, USA
Sandra Fonseca, CISA, CISM, CRISC, Metropolitan University (UMET), Puerto Rico
Ramaswami Karunanithi, CISA, CGEIT, CRISC, CA, CAMS, CBCI, CFE, CFSA, CGAP, CGMA, CIA, CMA, CPA, CRMA, CSCA, FCS, PMP, Prince2 Practitioner, New South Wales Government, Australia
Shruti Shrikant Kulkarni, CISA, CRISC, CISSP, CPISI, CCSK, ITIL V3 Expert, Monitise Group Ltd, United Kingdom
Ravikumar Ramachandran, CISA, CISM, CGEIT, CRISC, CAP, CEH, CFE, CHFI, CIA, CIMA-Adv.Dip.MA, CISSP-ISSAP, CRMA, ECSA, FCMA, PMP, SSCP, Hewlett-Packard India Sales Pvt.Ltd, India
Rahul Sharma, CISA, CISM, CRISC, CISSP, Bank of Montreal, Canada
Darron Sun, CISA, CRISC, CISSP, CMA, CRMA, Manulife, Hong Kong
Jan van Prooijen, CISA, CRISC, CISSP, HP Enterprise Security Services, The Netherlands
Jonathan Waldo, CISA, CRISC, Prime Therapeutics LLC, USA

ISACA has begun planning the seventh edition of the *CRISC™ Review Manual*. Volunteer participation drives the success of the manual. If you are interested in becoming a member of the select group of professionals involved in this global project, we want to hear from you. Please email us at *studymaterials@isaca.org*.

CRISC
Certified in Risk and Information Systems Control
An ISACA Certification

CRISC
Certified in Risk and Information Systems Control™
An ISACA® Certification

CRISC
Certified in Risk and Information Systems Control™
An ISACA® Certification

About This Manual

OVERVIEW

The *CRISC™ Review Manual 6th Edition* is a reference guide designed to assist candidates in preparing for the CRISC examination. **The manual is one source of preparation for the exam, but should not be thought of as the only source nor viewed as a comprehensive collection of all the information and experience that are required to pass the exam.** No single publication offers such coverage and detail.

As candidates read through the manual and encounter topics that are new to them or ones in which they feel their knowledge and experience are limited, additional references should be sought. The examination will be composed of questions testing the candidate's technical and practical knowledge, and ability to apply the knowledge (based on experience) in given situations.

ORGANIZATION OF THIS MANUAL

The *CRISC™ Review Manual 6th Edition* is divided into four chapters covering the CRISC domains tested on the exam in the percentages listed below:

Domain 1	IT Risk Identification	27 percent
Domain 2	IT Risk Assessment	28 percent
Domain 3	Risk Response and Mitigation	23 percent
Domain 4	Risk and Control Monitoring and Reporting	22 percent

Note: Each chapter defines the tasks that CRISC candidates are expected to know how to do and includes a series of knowledge statements required to perform those tasks. These constitute the current practices for the IT risk professional. **The detailed CRISC job practice can be viewed at *www.isaca.org/criscjobpractice*. The exam is based on these task and knowledge statements.**

The manual has been developed and organized to assist in the study of these areas. Exam candidates should evaluate their strengths, based on knowledge and experience, in each of these areas.

FORMAT OF THIS MANUAL

The first section of this manual introduces IT risk management.

Each of the four chapters of the *CRISC™ Review Manual 6th Edition* is divided into two sections for focused study.

Section One is an overview that provides:
- A definition for each domain
- Learning objectives for each domain
- A listing of the task and knowledge statements for each domain
- Sample self-assessment questions, answers and explanations
- Suggested resources for further study

Section Two consists of reference material and content that supports the knowledge necessary to perform the task statements. Material included is pertinent to CRISC candidates' knowledge and/or understanding when preparing for the CRISC certification exam.

The structure of the content includes numbering to identify the chapter where a topic is located and headings of the subsequent levels of topics addressed in the chapter (i.e., 1.3.1 Risk Culture is a subtopic of 1.3 Risk Culture and Communication in chapter 1). Relevant content in a subtopic is bolded for specific attention.

Understanding the material is a barometer of the candidate's knowledge, strengths and weaknesses, and is an indication of areas in which the candidate needs to seek reference sources over and above this manual. However, written material is not a substitute for experience. **CRISC exam questions will test the candidate's practical application of this knowledge.** The self-assessment questions in the first section of each chapter assist in understanding how a CRISC question could be presented on the CRISC exam and should not be used independently as a source of knowledge. Self-assessment questions should not be considered a measurement of the candidate's ability to answer questions correctly on the CRISC exam for that area. The questions are intended to familiarize the candidate with question structure, and may or may not be similar to questions that will appear on the actual examination. The reference material includes other publications that could be used to further acquire and better understand detailed information on the topics addressed in the manual.

A glossary is included at the end of the manual and contains terms that apply to the material included in the chapters. Also included are terms that apply to related areas not specifically discussed. The glossary is an extension of the text in the manual and can, therefore, be another indication of areas in which the candidate may need to seek additional references.

Although every effort is made to address the majority of information that candidates are expected to know, not all examination questions are necessarily covered in the manual, and candidates will need to rely on professional experience to provide the best answer.

Throughout the manual, "association" refers to ISACA, formerly known as Information Systems Audit and Control Association, and "institute" or "ITGI®" refers to the IT Governance Institute®. Also, please note that the manual has been written using standard American English.

Note: The *CRISC™ Review Manual 6th Edition* is a living document. As the field of IT-related business risk management in information systems controls evolve, the manual will be updated to reflect such changes. Further updates to this document before the date of the exam may be viewed at *www.isaca.org/studyaidupdates.*

EVALUATION OF THIS MANUAL

ISACA continuously monitors the swift and profound professional, technological and environmental advances affecting the IS risk profession. Recognizing these rapid advances, the *CRISC™ Review Manual* is updated annually.

To assist ISACA in keeping abreast of these advances, please take a moment to evaluate the *CRISC™ Review Manual 6th Edition*. Such feedback is valuable to fully serve the profession and future CRISC exam registrants.

To complete the evaluation on the web site, please go to *www.isaca.org/studyaidsevaluation*.

Thank you for your support and assistance.

ABOUT THE CRISC REVIEW QUESTIONS, ANSWERS & EXPLANATIONS MANUAL

Candidates may also wish to enhance their study and preparation for the exam by using the *CRISC™ Review Questions, Answers & Explanations Manual 4th Edition.*

The *CRISC™ Review Questions, Answers & Explanations Manual 4th Edition* consists of 500 multiple-choice study questions, answers and explanations arranged in the domains of the current CRISC job practice. The questions in this manual appeared in the *CRISC™ Review Questions, Answers & Explanations Manual 2015* and in the *CRISC™ Review Questions, Answers & Explanations Manual 2015 Supplement.*

Questions in this publication are representative of the types of questions that could appear on the exam and include explanations of the correct and incorrect answers. Questions are sorted by the CRISC domains and as a sample test. These publications are ideal for use in conjunction with the *CRISC™ Review Manual 6th Edition*. These manuals can be used as study sources throughout the study process or as part of a final review to determine where candidates may need additional study. It should be noted that these questions and suggested answers are provided as examples; they are not actual questions from the examination and may differ in content from those that actually appear on the exam.

ABOUT THE CRISC REVIEW QUESTIONS, ANSWERS & EXPLANATIONS DATABASE

Another study aid that is available is the CRISC™ Review Questions, Answers & Explanations Database – 12 Month Subscription. The online database consists of the 500 questions, answers and explanations included in the *CRISC™ Review Questions, Answers & Explanations Manual 4th Edition*. With this product, CRISC candidates can quickly identify their strengths and weaknesses by taking random sample exams of varying length and breaking the results down by domain. Sample exams also can be chosen by domain, allowing for concentrated study, one domain at a time, and other sorting features such as the omission of previous correctly answered questions are available.

> **Note:** When using the CRISC review materials to prepare for the exam, it should be noted that they cover a broad spectrum of IT-related business risk and IS control issues. **Again, candidates should not assume that reading these manuals and answering review questions will fully prepare them for the examination.** Since actual exam questions often relate to practical experiences, candidates should refer to their own experiences and other reference sources, and draw on the experiences of colleagues and others who have earned the CRISC designation.

Introduction to IT Risk Management

Introduction to IT Risk Management

The definition of risk has changed over time, and several definitions are currently in use. They ultimately convey the same meaning: risk is the combination of the probability of an event and its consequence. This definition is intentionally broad because risk is a business factor that has the potential to influence both positive and negative outcomes. In the context of IT, risk is often seen as an adverse factor that can threaten an organization's assets or otherwise cause harm. Several factors are considered when evaluating risk, including the mission of the organization, its assets, threat and vulnerability, likelihood, and consequences (also called impact). These terms will be further explored in each of the chapters in this review manual.

GOVERNANCE AND RISK MANAGEMENT

Governance is the accountability for protection of the assets of an organization. In a corporate structure, the directors of an organization (frequently organized as a board) are accountable for governance and entrust the senior management team with the responsibility to manage the day-to-day operations of the organization in alignment with the strategic mandates that the directors approve. Similar arrangements exist within cooperative and partnership-style organizations, although the names may differ.

Governance is applicable to all departments of the organization. It may take the form of financial accountability and oversight, operational effectiveness, legal and regulatory compliance, adoption of fair labor practices, social responsibility and governance of IT investment, operations, and control. Risk management is an important part of governance. Managers require accurate information to be able to correctly understand risk and address the circumstances that would indicate the need for risk mitigation.

Over the past decade, the term "governance" has moved to the forefront of business thinking in response to examples demonstrating the importance of good governance on one end of the spectrum and the global business mishaps derived from poor governance on the other. Corporate governance is the system by which organizations are evaluated, directed and controlled. By implication, the corporate governance of IT is the system by which the current and future use of IT is evaluated, directed and controlled. The objective of any governance system is to enable organizations to create value for their stakeholders or to promote value creation. Value creation, in turn, is comprised of benefits realization, risk optimization and resource optimization. Risk optimization is an essential part of any governance system and cannot be seen in isolation from benefits realization or resource optimization.

Governance answers four questions:
1. Are we doing the right things?
2. Are we doing them the right way?
3. Are we getting them done well?
4. Are we getting the benefits?

There is a clear distinction between governance and management. Management focuses on planning, building, running and monitoring activities in alignment with the direction set by the governance body to create value by achieving objectives. A well-managed organization subject to poor governance will create and execute clear, effective plans to attain objectives that do not create value. Similarly, risk management foresees the challenges to achieving objectives and attempts to lower the probabilities of negative outcomes occurring and/or their impacts if they do occur, but the effectiveness of risk management depends in large part on decisions made by managers responsible for risk governance.

Figure 0.1 provides an overview of the risk governance structure.

Figure 0.1—Risk Governance

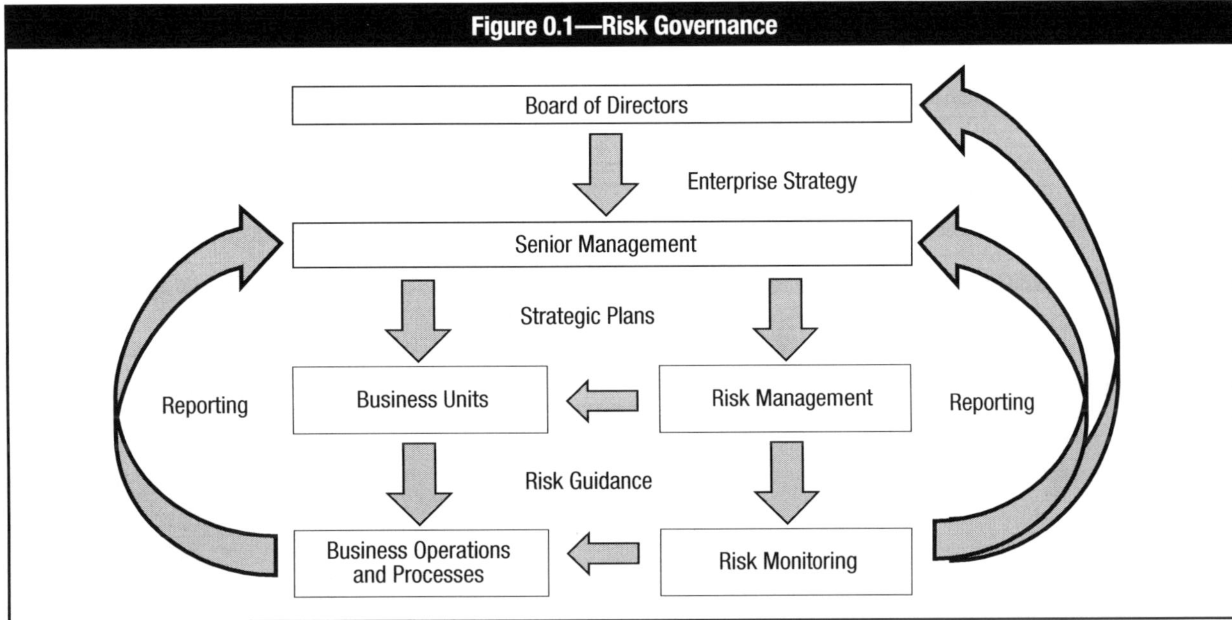

Effective risk governance helps ensure that risk management practices are embedded in the enterprise, enabling it to secure optimal risk-adjusted return.

Figure 0.2 describes the four main objectives of risk governance.

Figure 0.2—Risk Governance Objectives

Objective	Description
1. Establish and maintain a common risk view.	Effective risk governance establishes the common view of risk for the enterprise. This determines which controls are necessary to mitigate risk and how risk-based controls are integrated into business processes and information security. The risk governance function sets the tone of the business regarding how to determine an acceptable level of risk tolerance. Risk governance is a continuous life cycle that requires regular reporting and ongoing review. The risk governance function must oversee the operations of the risk management team.
2. Integrate risk management into the enterprise.	Integrating risk management into the enterprise enforces a holistic enterprise risk management (ERM) approach across the entire enterprise. It requires the integration of risk management into every department, function, system and geographic location. Understanding that risk in one department or system may pose an unacceptable risk to another department or system requires that all business processes be compliant with a baseline level of risk management. The objective of ERM is to establish the authority to require all business processes to undergo a risk analysis on a periodic basis or when there is a significant change to the internal or external environment.
3. Make risk-aware business decisions.	To make risk-aware business decisions, the risk governance function must consider the full range of opportunities and consequences of each such decision and its impact on the enterprise, society and the environment.
4. Ensure that risk management controls are implemented and operating correctly.	Governance requires oversight and due diligence to ensure that the enterprise is following up on the implementation and monitoring of controls to ensure that the controls are effective to mitigate risk and protect organizational assets.

THE CONTEXT OF IT RISK MANAGEMENT

Risk management is defined as the coordinated activities to direct and control an enterprise with regard to risk. In simple terms, risk can be viewed as a challenge to achieving objectives, and risk management as the activity undertaken to predict challenges and lower their chances of occurring and/or their impact. Effective risk management can also assist in maximizing opportunities, and the risk practitioner should keep this upside/downside duality of risk in mind. For example, a risk decision might take the form of potential benefits that may accrue if opportunities are taken versus missed benefits if those same opportunities are foregone.

The dual nature of risk is a result of its use in different contexts by business and IT, and it is not always easy to draw the distinction. *International Organization for Standardization (ISO) 31000:2009 – Risk Management Principles and Guidelines* calls risk "the effect of uncertainty on objectives. An effect is a deviation from the expected—positive and/or negative." However, *ISO/International Electrotechnical Commission (IEC) 27005:2011 – Information technology – Security techniques – Information security risk management* regards risk solely from a negative angle, stating "information security risk is the potential that a given threat will exploit vulnerabilities of an asset or group of assets and thereby cause harm to the organization." Risk practitioners who are able to effectively view risk from both perspectives are likely to find that they can more easily discuss risk with business and IT professionals without causing confusion.

Risk management starts with understanding the organization, but the risk practitioner should bear in mind that the organization is heavily influenced by the environment, or context, in which it operates. Assessing an organization's context includes evaluating the intent and capability of threats; the relative value of assets or resources and the trust that must be placed in them; and the presence and extent of vulnerabilities that might be exploited to intercept, interrupt, modify or fabricate data in information assets. Other factors that must be considered include:
- Dependency of the organization on a supply chain, especially one based in another geographic region of the world or reliant upon just-in-time delivery
- The influences of financing, debt and partners or substantial stakeholders
- Vulnerability to changes in economic or political conditions
- Changes to market trends and patterns
- Emergence of new competition
- Impact of new legislation
- Existence of potential natural disaster
- Constraints caused by legacy systems and antiquated technology
- Strained labor relations and inflexible management

The strategy of the organization will drive the individual lines of business that make up the organization, and each line of business will develop information systems that support its business function. **Figure 0.3** illustrates how IT risk relates to overall risk of the organization.

Figure 0.3—IT Risk in the Risk Hierarchy

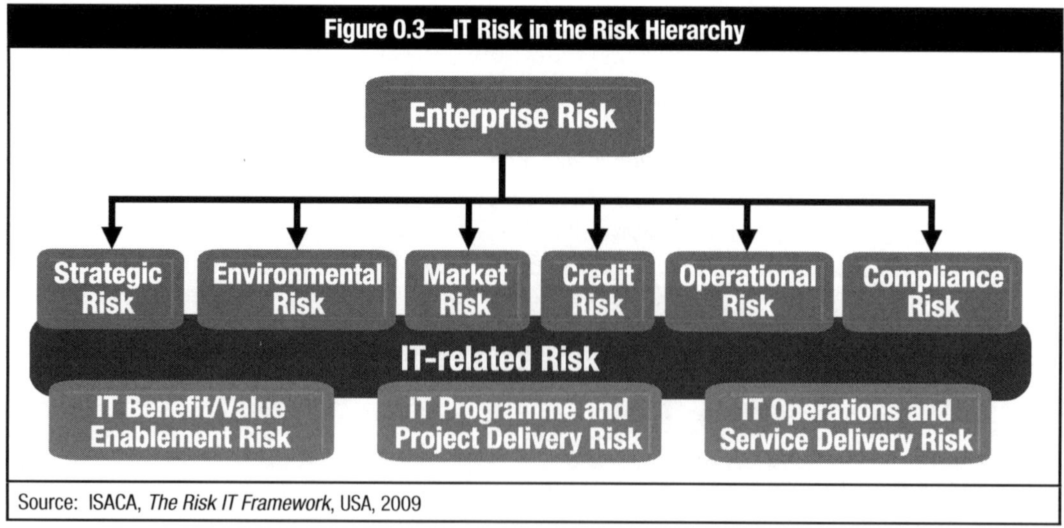

Source: ISACA, *The Risk IT Framework*, USA, 2009

Examples of specific risk that corresponds to the categories shown in **figure 0.3** include the following:
- Strategic: Changes in customer preference or stakeholder preference, executive turnover
- Environmental: Pollution or disturbance of protected areas
- Market: Foreign-exchange rates, availability of commodities and raw materials
- Credit: Interest rates, callable loans, damage to assets for which the organization is an insurer
- Operational: Employee errors, fraud, theft
- Compliance: Failure to meet regulatory requirements, inaccurate documentation
- IT benefit/value enablement: Delivered projects do not create expected value
- IT program and project delivery: Projects are not delivered in a manner consistent with plans
- IT operations and service delivery: Delivered services fall short of service level agreements (SLAs)

Risk is an influencing factor and must be evaluated at all levels of the organization—the strategic level, the business unit level and the information systems level. A properly managed risk framework addresses the impact of risk at all levels and describes how risk at one level may affect the other levels as well.

There are several key parts of an IT risk management program. Different risk management methodologies use slightly different terms to describe the components of IT risk management. The CRISC candidate is not expected to be familiar with the details of each methodology, but should be familiar with the general concepts and process flows related to IT risk management.

IT risk management is the implementation of a risk strategy that reflects the culture, appetite and tolerance levels of organizational management; considers technology and budgets; and addresses the requirements of regulation and compliance. An effective IT risk management strategy is critical to an organization's ability to effectively and efficiently execute its overall business strategy.

IT risk management is a cyclical process. The first step in the IT risk management process is the identification of IT risk, which includes determining the risk context and risk framework, and the process of identifying and documenting risk. The risk identification effort should result in the listing and documentation of risk, which serves as the input for the next phase of the process, IT risk assessment. The effort to assess and prioritize risk provides management with the data needed for risk response and mitigation, the third phase of the cycle, which seeks and implements cost-effective ways to address the risk that has been identified and assessed. The final phase is risk and control monitoring and reporting, in which controls, risk management efforts and the current risk state are monitored and the results reported back to senior management. The process repeats as the risk environment changes, which may occur a as a result of internal or external factors.

Figure 0.4 illustrates the cyclical IT risk management process.

Figure 0.4—The IT Risk Management Life Cycle

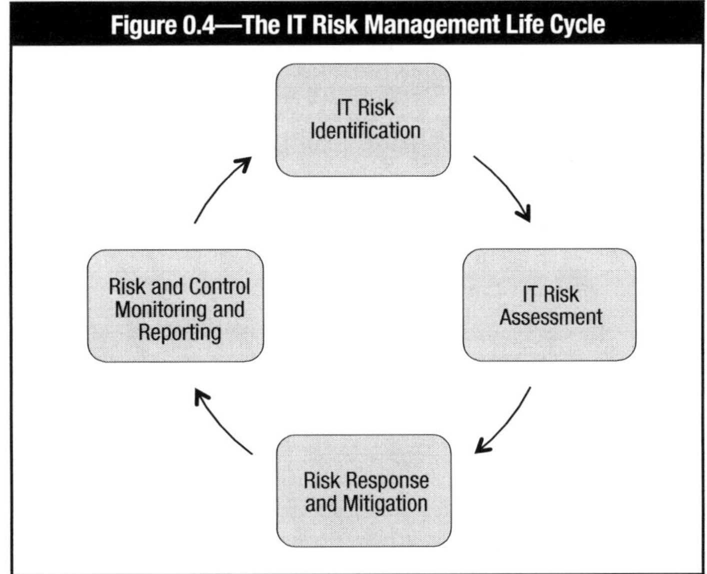

The IT risk management process is based on the complete cycle of all the elements. A failure to perform any one of the phases in a complete and thorough manner may result in deficiencies being carried forward that cause the overall process to be ineffective. As with all life cycles, the process continues with refinement, adaptation and a focus on continuous improvement and maturity. The more often the risk management life cycle is repeated, the more effective the IT risk management effort will be and the more consistency the organization will see in the results that it obtains.

Importance of IT Risk Management

IT risk management is important to the organization because of the tangible benefits that it delivers, including the following:

- Better oversight of organizational assets
- Reduced or minimized loss
- Identification of threats, vulnerabilities and consequences on a proactive basis
- Prioritization of risk response efforts to match organizational goals and priorities
- A more holistic basis for and approach to legal and regulatory compliance
- Increased likelihood of project success
- Improved performance, leading to greater stakeholder confidence
- Creation of a risk-aware culture with less reliance on specialists
- Better incident and business continuity management
- Improved controls with better monitoring and reporting
- Improved decision making as a result of expanded access to accurate, timely information
- An increased ability to meet business objectives and create value

KEY CONCEPTS OF RISK

Risk can be discussed in quantitative or qualitative terms, and the specific definitions of risk vary from source to source. However, the fundamental nature of risk is that it addresses the odds that some event will happen (probability or likelihood) and what it would mean for the organization if that event did happen (consequences). Early attempts to define risk observed that the probability of something happening was a combination of two things: whether something was attempted (threat) and whether the target of the attempt was susceptible to what was tried (vulnerability).

As the study of risk matured, risk practitioners began to distinguish between delineation of the consequences and the extent to which those consequences affected the value-creation activities of the organization (impact). It is now common to distinguish between different types of threats, to evaluate them on the basis of specific organizational assets against which they may be directed and to assess those assets in terms of their individual weaknesses (vulnerabilities) that might be exploited to create consequences for the assets. When viewed from the perspective of how these assets are used within the organization, it becomes possible to quantify impact in terms of lost productivity and other specific measures of value, which is useful for two reasons:

1. It is easier for managers to set a dollar value of total losses that they are willing to incur (risk appetite) than it is to define what consequences are or are not acceptable in a dozen or more different areas of operations.
2. Knowing the potential losses associated with risk provides a basis for deciding how to respond to risk that is beyond acceptable levels because it does not make sense to spend more to respond to a risk than the risk itself presents in terms of the cost of impact.

> Example of Risk:
>
> Consider a house in a dry, wooded area; fire is a threat regardless of the building material used to construct the house (i.e., wood or brick). The probability of a fire starting in the wooded area is distinct from the probability of the house burning down. For the second instance, we take into consideration the building material. A wooden house is more vulnerable to fire; a brick house is not. For the same threat (a fire starting), the likelihood of impact is, therefore, different depending on the vulnerability.
>
> Next, the destruction of the house is a potential consequence. If the house is occupied, the impact is temporary homelessness for those who live there, which imposes the immediate costs of temporary lodging and a replacement wardrobe. Under those circumstances, it makes sense to take precautions sufficient to address this impact, such as insuring the home against fire or putting in a fire-suppression system, but it is not reasonable to hire a full-time fire crew to watch the house on a daily basis because the cost of the fire crew would exceed the cost of the impact. However, if the house were uninhabited and condemned, the consequences would have no negative impact, and no precautions may be necessary.

The key concepts of risk are discussed in various contexts throughout this review manual. It is common for people who lack strong understanding of these terms to use them interchangeably, but doing so can create confusion, impede successful risk management and cast doubt on competence. The risk practitioner should ensure that he/she spends enough time studying them to gain a basic, reliable understanding of the different terms and how they relate to one another.

RISK IN RELATION TO OTHER BUSINESS FUNCTIONS

Risk is a critical part of business. Unless a business is willing to take a risk, it will not be able to realize the benefits associated with risk. However, taking too much risk may lead to increased likelihood of failure of the business and loss of investment. Senior management is responsible for setting the risk appetite for the organization—a clear statement of how much risk to take and what opportunities to forego.

The risk practitioner is primarily concerned with IT risk, which is a subset of business risk. This requires the risk practitioner to understand the risk culture of an organization and use it to drive or inform the IT risk strategy. The business does not exist so that the organization can have an IT department; the IT department exists to help the business meet its mission and goals. When calculating IT risk, the risk practitioner must be careful not to calculate risk solely from the perspective of the impact of the risk on IT and to ensure that both the technical and nontechnical elements of risk have been considered. An IT system failure has an impact on the IT department, but it may have a much greater impact on the business supported by the IT system than on IT alone.

In addition to the relationship between business risk and IT risk, the risk practitioner should also be familiar with risk-related business functions such as business continuity, audit, information security, controls, projects and change management.

Risk and Business Continuity

IT risk management is closely linked with business continuity. The business function is concerned with the preservation of critical business functions and the ability of the organization to survive an adverse event that may impact the ability of the organization to meet its mission and goals. Through risk management, the organization attempts to reduce all IT risk to an acceptable level. Although the controls and efforts of IT risk management may not prevent a failure, the risk practitioner works with the incident management and business continuity teams to identify possible threats and put in place the mechanism to detect, contain and recover from an adverse event if it should happen. If the business continuity plan (BCP) is inadequate or inaccurate, the organization may not meet its goals for recovery after an incident.

Risk and Audit

The audit function is an important part of corporate governance that provides management with assurance regarding the effectiveness of the control framework, IT risk management program and compliance. In a world of increasing legislation, government oversight and media scrutiny, organizations must diligently demonstrate an adequate control environment and risk management. For that reason, audits of information systems (IS) should be conducted by objective, skilled and independent personnel able to assess risk, identify vulnerabilities, document findings and provide recommendations on how to address audit issues.

An audit is a methodical and structured review that requires competence and knowledge in the subject matter of the area being audited. If the IS auditor is not familiar with the technology being used, the significance of operating conditions or the requirements of the organization, the audit may be inaccurate and provide limited value. IS audits must also be independent. Senior management is often involved in the creation of the IS audit plan, and in a situation where a particular manager is involved in inappropriate activity, he/she may restrict the ability of IS audit to perform their duties effectively. Even when there is no wrongdoing, the appearance of partiality may cast doubt on the results of an audit, creating less value than expected. The risk practitioner should review the relationship between the IS auditor and the area being audited to ensure that there is no conflict of interests.

Risk and Information Security

IT risk management drives the selection of controls and justifies their initial and continued operation. If the IT risk management activity is not conducted properly, information security controls are almost certain to be incorrectly designed, poorly implemented and improperly operated. Every control should be traceable back to a specific IT risk that the control is designed to mitigate, and the risk practitioner should be able to demonstrate the purpose of each control and explain the reasoning behind its selection.

Control Risk

A control is chosen to mitigate a risk, but if the control is not operating correctly then the control may not prevent a failure or compromise. The selection of the wrong control, the incorrect configuration of the control, the improper operation of the control, the failure to monitor and review the control, or the inadequacy of the control to address new threats may each introduce the risk of control failure.

Project Risk

Many projects fail; in fact, numerous studies of IT projects have indicated that a majority of IT projects could be considered failures. Failure of a particular project may be defined by it going over its allotted budget or the allotted time scheduled or if it does not deliver what it promised. A project may also be deemed a failure if it delivered what it promised but the deliverables did not meet customer needs and expectations. The failure of an IT project may pose a significant risk to an organization, manifesting as lost market share, failure to seize new opportunities or other adverse impacts on customers, shareholders and staff. Identifying the risk associated with a project and successfully managing that risk is very likely to result in higher levels of project success and stakeholder satisfaction.

Change Risk

Risk is not static. Changes in technology, regulations, business processes, functionality, architecture, users and other variables that affect the business and technical environments of the organization may affect the levels of risk associated with systems in operation. The risk level of a particular system may also change because of intentional changes to its configuration or architecture that result in the controls that were originally effective as designed becoming ineffective. The risk practitioner is tasked with managing risk on a continuous basis, which means that he or she must stay aware of emerging risk that may be associated with new threats, new technologies, changes in culture, and alterations in legislation and/or regulation. All of these changes may affect the risk posture of the organization and result in a new level of risk not adequately addressed in earlier risk identification efforts.

IT RISK MANAGEMENT GOOD PRACTICES

The process of IT risk management is most reliably effective when it follows a structured methodology based on good practices and a desire to seek continuous improvement. The risk practitioner should begin a new risk management effort by reviewing current practices of the organization in the identification, assessment, response, monitoring and reporting of risk. On the basis of this initial evaluation, the risk practitioner can gain valuable insight into how the organization views risk management and identify areas in which the current program may incorporate or deviate from recognized good practices, which can facilitate the development of a consistent program.

Where good practices are not already in place, the risk practitioner may find it beneficial to either formally adopt or informally draw upon one or more well-established standards or frameworks, which can help to ensure that the risk management program is complete and authoritative. Examples of standards and frameworks that may be useful sources of good practices include:

- *COBIT® 5 for Risk*
- Committee of Sponsoring Organizations of the Treadway Commission (COSO) Enterprise Risk Management – Integrated Framework
- Operationally Critical Threat, Asset and Vulnerability Evaluation (OCTAVE)
- International Organization for Standardization (ISO)
  - *ISO/IEC 27005:2011 – Information technology – Security techniques – Information security risk management*
  - *ISO 31000:2009 – Risk Management Principles and Guidelines*
- US National Institute of Standards and Technology Special Publications (NIST SPs)
  - SP 800-30 Revision 1: *Guide for Conducting Risk Assessments*
  - SP 800-39: *Managing Information Security Risk*

Section 1.7.2 Risk Identification and Classification Standards and Frameworks provides further information on each specific standard and framework.

The IT risk management program should be:

- Comprehensive (thorough, detailed)
- Complete (carried through to the end)
- Auditable (reviewable by an independent third party)
- Justifiable (based on sound reasoning)
- Compliant (with policy, laws and/or regulations)
- Monitored (subject to review and accountability)
- Enforced (consistent, mandated and required)
- Up to date (current with changing business processes, technologies and laws)
- Managed (adequately resourced, with oversight and support)

Good practices in IT risk management, like all other good practices, are subject to change and improvement. The risk practitioner should be aware of these changes and take steps to ensure that the organization adopts practices that are informed by the best available industry knowledge and experience, while keeping in mind that any standard or framework may need to be tailored to support the particular goals of the organization.

SUMMARY

This section provided an overview of the areas of IT risk that may be of interest or importance to the risk practitioner. There are many variables that a risk practitioner must consider in making recommendations, and the most successful, reliable IT risk management efforts tend to be those based on an enterprise risk perspective gathered by means of a structured methodology that provides timely, accurate information. The risk practitioner creates values for the organization by promoting and adhering to a well-developed, consistent IT risk management program.

ENDNOTES

[1] McManus, John; Trevor Wood-Harper; "A Study in Project Failure," BCS, The Chartered Institute for IT, June 2008, *www.bcs.org/content/ConWebDoc/19584*

Page intentionally left blank

Chapter 1: IT Risk Identification

Section One: Overview

Section Two: Content

Section One: Overview

DOMAIN DEFINITION

Identify the universe of IT risk to contribute to the execution of the IT risk management strategy in support of business objectives and in alignment with the enterprise risk management (ERM) strategy.

LEARNING OBJECTIVES

The objective of this domain is to ensure that the CRISC candidate has the knowledge necessary to:
- Identify relevant standards, frameworks and practices
- Apply risk identification techniques
- Distinguish between threats and vulnerabilities
- Identify relevant stakeholders
- Discuss risk scenario development tools and techniques
- Explain the meaning of key risk management concepts, including risk appetite and risk tolerance
- Describe the key elements of a risk register
- Contribute to the creation of a risk awareness program

CRISC EXAM REFERENCE

This domain represents 27 percent of the CRISC exam (approximately 41 questions).

TASK AND KNOWLEDGE STATEMENTS

TASKS

There are seven tasks within this domain that a CRISC candidate must know how to perform. These relate to IT risk identification.

T1.1 Collect and review information, including existing documentation, regarding the organization's internal and external business and IT environments to identify potential impacts of IT risk to the organization's business objectives and operations.

T1.2 Identify potential threats and vulnerabilities to the organization's people, processes and technology to enable IT risk analysis.

T1.3 Develop a comprehensive set of IT risk scenarios based on available information to determine the potential impact to business objectives and operations.

T1.4 Identify key stakeholders for IT risk scenarios to help establish accountability.

T1.5 Establish an IT risk register to help ensure that identified IT risk scenarios are accounted for and incorporated into the enterprisewide risk profile.

T1.6 Identify risk appetite and tolerance defined by senior leadership and key stakeholders to ensure alignment with business objectives.

T1.7 Collaborate in the development of a risk awareness program, and conduct training to ensure that stakeholders understand risk and to promote a risk-aware culture.

KNOWLEDGE STATEMENTS

The CRISC candidate should be familiar with the task statements relevant to each domain in the CRISC job practice. The tasks are supported by 41 knowledge statements that delineate each of the areas in which the risk practitioner must have a good understanding in order to perform the tasks. Many knowledge statements support tasks that cross domains.

The CRISC candidate should have knowledge of:
1. Laws, regulations, standards and compliance requirements
2. Industry trends and emerging technologies
3. Enterprise systems architecture (e.g., platforms, networks, applications, databases and operating systems)
4. Business goals and objectives
5. Contractual requirements with customers and third-party service providers

6. Threats and vulnerabilities related to:
   6.1 Business processes and initiatives
   6.2 Third-party management
   6.3 Data management
   6.4 Hardware, software and appliances
   6.5 The system development life cycle (SDLC)
   6.6 Project and program management
   6.7 Business continuity and disaster recovery management (DRM)
   6.8 Management of IT operations
   6.9 Emerging technologies
7. Methods to identify risk
8. Risk scenario development tools and techniques
9. Risk identification and classification standards, and frameworks
10. Risk events/incident concepts (e.g., contributing conditions, lessons learned, loss result)
11. Elements of a risk register
12. Risk appetite and tolerance
13. Risk analysis methodologies (quantitative and qualitative)
14. Organizational structures
15. Organizational culture, ethics and behavior
16. Organizational assets (e.g., people, technology, data, trademarks, intellectual property) and business processes, including enterprise risk management (ERM)
17. Organizational policies and standards
18. Business process review tools and techniques
19. Analysis techniques (e.g., root cause, gap, cost-benefit, return on investment [ROI])
20. Capability assessment models and improvement techniques and strategies
21. Data analysis, validation and aggregation techniques (e.g., trend analysis, modeling)
22. Data collection and extraction tools and techniques
23. Principles of risk and control ownership
24. Characteristics of inherent and residual risk
25. Exception management practices
26. Risk assessment standards, frameworks and techniques
27. Risk response options (i.e., accept, mitigate, avoid, transfer) and criteria for selection
28. Information security concepts and principles, including confidentiality, integrity and availability of information
29. Systems control design and implementation, including testing methodologies and practices
30. The impact of emerging technologies on design and implementation of controls
31. Requirements, principles, and practices for educating and training on risk and control activities
32. Key risk indicators (KRIs)
33. Risk monitoring standards and frameworks
34. Risk monitoring tools and techniques
35. Risk reporting tools and techniques
36. IT risk management best practices
37. Key performance indicator (KPIs)
38. Control types, standards, and frameworks
39. Control monitoring and reporting tools and techniques
40. Control assessment types (e.g., self-assessments, audits, vulnerability assessments, penetration tests, third-party assurance)
41. Control activities, objectives, practices and metrics related to:
   41.1 Business processes
   41.2 Information security, including technology certification and accreditation practices
   41.3 Third-party management, including service delivery
   41.4 Data management
   41.5 The system development life cycle (SDLC)
   41.6 Project and program management
   41.7 Business continuity and disaster recovery management (DRM)
   41.8 IT operations management
   41.9 The information systems architecture (e.g., platforms, networks, applications, databases and operating systems)

SELF-ASSESSMENT QUESTIONS

CRISC self-assessment questions support the content in this manual and provide an understanding of the type and structure of questions that have typically appeared on the exam. Questions are written in a multiple-choice format and designed for one best answer. Each question has a stem (question) and four options (answer choices). The stem may be written in the form of a question or an incomplete statement. In some instances, a scenario or a description problem may also be included. These questions normally include a description of a situation and require the candidate to answer two or more questions based on the information provided. Many times a question will require the candidate to choose the **MOST** likely or **BEST** answer among the options provided.

In each case, the candidate must read the question carefully, eliminate known incorrect answers and then make the best choice possible. Knowing the format in which questions are asked, and how to study and gain knowledge of what will be tested, will help the candidate correctly answer the questions.

1-1 Which of the following business requirements **BEST** relates to the need for resilient business and information systems processes?

A. Effectiveness
B. Confidentiality
C. Integrity
D. Availability

1-2 Which of the following statements **BEST** describes the value of a risk register?

A. It captures the risk inventory.
B. It drives the risk response plan.
C. It is a risk reporting tool.
D. It lists internal risk and external risk.

1-3 Shortly after performing the annual review and revision of corporate policies, a risk practitioner becomes aware that a new law may affect security requirements for the human resources system. The risk practitioner should:

A. analyze in detail how the law may affect the enterprise.
B. ensure that necessary adjustments are implemented during the next review cycle.
C. initiate an *ad hoc* revision of the corporate policy.
D. notify the system custodian to implement changes.

1-4 An information system that processes weather forecasts for public consumption is **MOST** likely to place its highest priority on:

A. nonrepudiation.
B. confidentiality.
C. integrity.
D. availability.

1-5 Which of the following choices provides the **BEST** view of risk management?

A. An interdisciplinary team
B. A third-party risk assessment service provider
C. The enterprise's IT department
D. The enterprise's internal compliance department

1-6 Which of the following choices is a **PRIMARY** consideration when developing an IT risk awareness program?

A. Why technology risk is owned by IT
B. How technology risk can impact each attendee's area of business
C. How business process owners can transfer technology risk
D. Why technology risk is more difficult to manage compared to other risk

1-7 It is **MOST** important that risk appetite is aligned with business objectives to ensure that:

A. resources are directed toward areas of low risk tolerance.
B. major risk is identified and eliminated.
C. IT and business goals are aligned.
D. the risk strategy is adequately communicated.

1-8 Weak passwords and transmission over unprotected communication lines are examples of:

A. vulnerabilities.
B. threats.
C. probabilities.
D. impacts.

ANSWERS TO SELF-ASSESSMENT QUESTIONS

Correct answers are shown in **bold**.

1-1 A. Effectiveness deals with information being relevant and pertinent to the business process as well as being delivered in a timely, correct, consistent and usable manner. While the lack of system resilience can in some cases affect effectiveness, resilience is more closely linked to the business information requirement of availability.

B. Confidentiality deals with the protection of sensitive information from unauthorized disclosure. While the lack of system resilience can in some cases affect data confidentiality, resilience is more closely linked to the business information requirement of availability.

C. Integrity relates to the accuracy and completeness of information as well as to its validity in accordance with business values and expectations. While the lack of system resilience can in some cases affect data integrity, resilience is more closely linked to the business information requirement of availability.

D. Availability relates to information being available when required by the business process—now and in the future. Resilience is the ability to provide and maintain an acceptable level of service during disasters or when facing operational challenges.

1-2 A. A risk register is used to provide detailed information on each identified risk such as risk owner, details of the scenario and assumptions, affected stakeholders, causes/indicators, information on the detailed scores (i.e., risk ratings) on the risk analysis, and detailed information on the risk response (e.g., action owner and the risk response status, time frame for action, related projects, and risk tolerance level). These components can also be defined as the risk universe.

B. Risk registers serve as the main reference for all risk-related information, supporting risk-related decisions such as risk response activities and their prioritization.

C. Risk register data are utilized to generate management reports, but are not in themselves a risk reporting tool.

D. The risk register tracks all internal and external risk, the quality and quantity of the controls, and the likelihood and impact of the risk.

1-3 **A. Assessing how the law may affect the enterprise is the best course of action. The analysis must also determine whether existing controls already address the new requirements.**

B. Ensuring that necessary adjustments are implemented during the next review cycle is not the best answer, particularly when the law does affect the enterprise. While an annual review cycle may be sufficient in general, significant changes in the internal or external environment should trigger an *ad hoc* reassessment.

C. Corporate policy should be developed in a systematic and deliberate manner. An *ad hoc* amendment to the corporate policy is not warranted and may create risk rather than reducing it.

D. Notifying the system custodian to implement changes is inappropriate. Changes to the system should be implemented only after approval by the process owner.

1-4 A. Nonrepudiation refers to the ability to verifiably prove the originator of data, which is unlikely to be of importance for weather forecasts that are rendered accurately.

B. Keeping data confidential would be at odds with the business purpose of a system designed to provide data for public use.

C. A system that delivers weather forecasts is likely to place its highest priority on the integrity of the data. The risk practitioner should keep in mind that whether a forecast turns out to be accurate in its prediction is distinct from whether the data was accurately represented.

D. Availability of data is likely to be a lower priority for a weather-forecasting system than the accuracy with which the data is presented.

1-5 **A. Having an interdisciplinary team contribute to risk management ensures that all areas are adequately considered and included in the risk assessment processes to support an enterprise view of risk.**
B. Engaging a third party to perform a risk assessment may provide additional expertise to conduct the risk assessment; but without internal knowledge, it will be difficult to assess the adequacy of the risk assessment performed.
C. A risk assessment performed by the enterprise's IT department is unlikely to reflect the view of the entire enterprise.
D. The internal compliance department ensures the implementation of risk responses based on the requirement of management. It generally does not take an active part in implementing risk responses for items that do not have regulatory implications.

1-6 A. IT does not own technology risk. An appropriate topic of IT risk awareness training may be the fact that many types of IT risk are owned by the business. One example may be the risk of employees exploiting insufficient segregation of duties (SoD) within an enterprise resource planning (ERP) system.
B. Stakeholders must understand how the IT-related risk impacts the overall business.
C. Transferring risk is not of primary consideration in developing a risk awareness program. It is a part of the risk response process.
D. Technology risk may or may not be more difficult to manage than other types of risk. Although this is important from an awareness point of view, it is not as primary as understanding the impact in the area of business.

1-7 **A. Risk appetite is the amount of risk that an enterprise is willing to take on in pursuit of value. Aligning it with business objectives allows an enterprise to evaluate and deploy valuable resources toward those objectives where the risk tolerance (for loss) is low.**
B. There is no link between aligning risk appetite with business objectives and identification and elimination of major risk, and although risk can typically be reduced to an acceptable level using various risk response options, its elimination is rarely cost-effective even when it is possible.
C. Alignment of risk appetite with business objectives does converge IT and business goals to a point, but alignment is not limited to these two areas. Other areas include organizational, strategic and financial objectives, among other objectives.
D. Communication of the risk strategy does not depend on aligning risk appetite with business objectives.

1-8 **A. Vulnerabilities represent characteristics of information resources that may be exploited by a threat.**
B. Threats are circumstances or events with the potential to cause harm to information resources.
C. Probabilities represent the likelihood of the occurrence of a threat.
D. Impacts represent the outcome or result of a threat exploiting a vulnerability.

Note: For more self-assessment questions, you may also want to obtain a copy of the *CRISC™ Review Questions, Answers & Explanations Manual 4th Edition* or the Database, which each consist of 500 multiple-choice study questions, answers and explanations

SUGGESTED RESOURCES FOR FURTHER STUDY

In addition to the resources cited throughout this manual, the following resources are suggested for further study in this domain (publications in **bold** are stocked in the ISACA Bookstore):

Abkowitz, Mark D.; *Operational Risk Management: A Case Study Approach to Effective Planning and Response*, John Wiley & Sons, USA, 2008

The Association of Insurance and Risk Managers (Airmic); The Public Risk Management Association (Alarm); The Institute of Risk Management (IRM); *A Structured Approach to Enterprise Risk Management (ERM) and the Requirements of ISO 31000*, United Kingdom, 2010

Fraser, John; Betty J. Simkins (Eds.); *Enterprise Risk Management: Today's Leading Research and Best Practices for Tomorrow's Executives*, John Wiley & Sons, USA, 2010

Hillson, David; Ruth Murray-Webster; *Understanding and Managing Risk Attitude*, Gower Publishing Company, United Kingdom, 2005

The Institute of Internal Auditors—Australia; *HB 158-2010 Delivering assurance based on ISO 31000:2009—Risk management—Principles and guidelines*, Standards Australia GPO, 2010, Australia

The Institute of Risk Management (IRM); *Risk Appetite and Tolerance*, United Kingdom, 2011

International Organization for Standardization (ISO)/International Electrotechnical Commission (IEC); *ISO/IEC 27001:2013 Information Technology—Security Techniques—Information Security Management Systems—Requirements*, Switzerland, 2013

ISO/IEC; *ISO/IEC 27005:2011—Information technology—Security techniques—Information security risk management*, Switzerland, 2011

ISO/IEC; *ISO 31000:2009 Risk management—Principles and guidelines*, Switzerland, 2009

ISACA, *COBIT® 5 for Risk*, USA, 2013, *www.isaca.org/cobit*

ISACA, *Risk Scenarios: Using COBIT 5 for Risk*, USA, 2014

Moeller, Robert, R.; *COSO Enterprise Risk Management: Establishing Effective Governance, Risk, and Compliance Processes, 2nd Edition*, John Wiley & Sons, USA, 2011

National Institute of Standards and Technology (NIST); *NIST Special Publication 800-30 Revision 1: Guide for Conducting Risk Assessments*, USA, 2012

NIST; *NIST Special Publication 800-39: Managing Information Security Risk*, USA, 2011

Pickett, K. H. Spencer; *Auditing the Risk Management Process*, John Wiley & Sons, USA, 2013

The Risk Management Society (RIMS); The Institute of Internal Auditors (IIA); *Risk Management and Internal Audit: Forging a Collaborative Alliance*, USA, 2012

Rittenberg, Larry; Frank Martens; *Enterprise Risk Management: Understanding and Communicating Risk Appetite*, Committee of Sponsoring Organizations of the Treadway Commission (COSO), USA, 2012

Sobel, Paul J.; *Auditor's Risk Management Guide: Integrating Auditing and ERM*, Wolters Kluwer/CCH, USA, 2013

Walker, Paul L.; William G. Shenkir; Thomas L. Barton; *Improving Board Risk Oversight Through Best Practices*, The Institute of Internal Auditors Research Foundation, USA, 2011

Section Two: Content

1.0 OVERVIEW

Risk identification is the process for discovering, recognizing and documenting the risk an organization faces, and it is the first of the four processes that operate within the risk management life cycle. As shown in **figure 1.1**, identification is often initiated by the monitoring and reporting that occur in the last process of the life cycle, which may alert the risk practitioner to changes in the risk environment.

Figure 1.1—The IT Risk Management Life Cycle

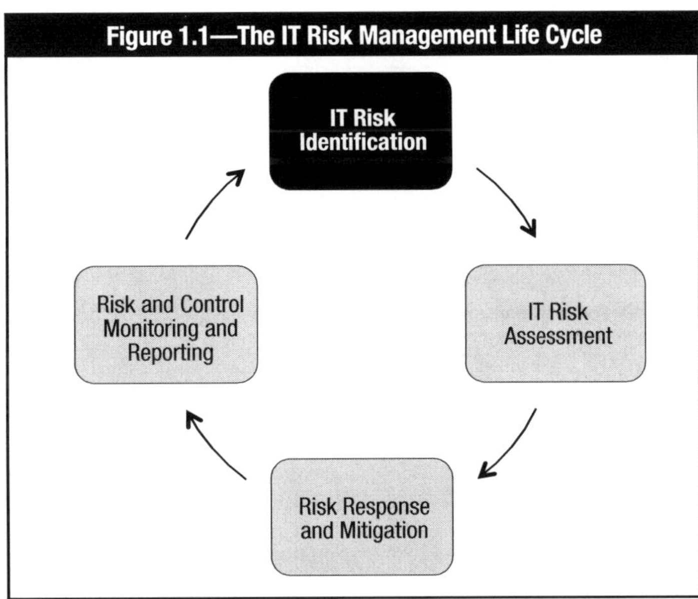

Risk identification is important because only risk that is identified can be assessed and subjected to appropriate responses. When the organization fails to identify a risk, it is left entirely outside of the strategic planning process used by senior management to promote value creation. The risk practitioner should work closely with business process owners to ensure that he or she has an understanding of the ways in which the organization operates that is broad enough to include not only the organization itself but also its external dependencies and assumptions, such as availability of contract labor or delivery of materials on a just-in-time basis.

1.1 RISK CAPACITY, RISK APPETITE AND RISK TOLERANCE

Every organization has a particular risk capacity, defined as the objective amount of loss an enterprise can tolerate without its continued existence being called into question. Subject to the absolute maximum imposed by this risk capacity, the owners or board of directors of an organization set the risk appetite for the organization. Risk appetite is defined as the amount of risk, on a broad level, that an entity is willing to accept in pursuit of its mission. In some cases, setting the risk appetite may be delegated by the board of directors to senior management as part of strategic planning.

Some benefits of defining risk capacity and risk appetite at the enterprise level include:
- Supporting and providing evidence of the risk-based decision-making processes
- Supporting the understanding of how each component of the enterprise contributes to the overall risk profile
- Showing how different resource allocation strategies can add to or lessen the burden of risk by simulating different risk response options
- Supporting the prioritization and approval process of risk response actions through risk budgets
- Identifying specific areas where a risk response should be made

Risk appetite is translated into a number of standards and policies to contain the risk level within the boundaries set by the risk appetite.[1] These boundaries need to be regularly adjusted or confirmed. Within these boundaries, risk may be accepted, a formal and explicit process that affirms that the risk requires and warrants no additional response by the organization as long as it and the risk environment stay substantially the same and accountability for the risk is assigned to a specific owner.

Risk acceptance generally should not exceed the risk appetite of the organization, but it must not exceed the risk capacity (which would threaten the continued existence of the organization). Risk tolerance levels are deviations from risk appetite, which are not desirable but are known to be sufficiently below the risk capacity that acceptance of risk is still possible when there is compelling business need and other options are too costly. Risk tolerance may be defined using IT process metrics or adherence to defined IT procedures and policies, which are a translation of the IT goals that need to be achieved. Like risk appetite, risk tolerance is defined at the enterprise level and reflected in the policies created by senior management. Exceptions can be tolerated at lower levels of the enterprise as long as the overall exposure does not exceed the risk appetite at the enterprise level.[2]

Note: Risk tolerance is defined as the acceptable level of variation that management is willing to allow for any particular risk as the enterprise pursues its objectives. The interpretation of the ISACA definition is that while management has an official acceptance level of one value, they may accept a slight deviation from that level. An example of tolerance is a situation where the speed limit on a highway is 65 miles/hour, but a police officer may allow a person to travel up to 70 miles/hour before issuing a ticket.

Risk appetite and tolerance should be defined and approved by senior management and clearly communicated to all stakeholders, and a process should be in place to review and approve any exceptions.

As with all risk, risk appetite and tolerance change over time, and factors such as new technology, organizational restructuring, or changes in business strategy may require the organization to reassess its risk portfolio and reconfirm its risk appetite.[3]

Creating a risk-aware culture is an important aspect of monitoring risk tolerances. This may be a charge of senior management through revisiting and reinforcing the risk appetite and holding the proper stakeholders accountable for implementing risk management with the organization's risk appetite.

According to the Committee of Sponsoring Organizations of the Treadway Commission (COSO), this is best accomplished when the organization communicates the risk appetite and risk tolerance, as discussed previously, with the following outcomes:[4]

- Consistent implementation across units
- Effective monitoring and communication of risk and changes in risk appetite
- Consistent understanding of risk appetite and related tolerances for each organizational unit
- Consistency between risk appetite, objectives and relevant reward systems

1.2 RISK CULTURE AND COMMUNICATION

Risk management is a core part of corporate governance. The governance of the assets and mission of the organization is reflected in the ways in which the organization seeks to protect its assets and attain its goals, and risk is a factor that may lead to failure or a loss of asset value. Understanding risk includes understanding the goals, objectives, values and ethics of the organization.

1.2.1 Risk Culture

Consciously or unconsciously, senior management develops and conveys a particular level of willingness to embrace, cautiously accept or avoid risk; this is called the risk culture of the organization.

The best indicator of the organization's risk culture is how it handles risk. Risk culture reflects a balance between weighing the negative, positive and regulatory elements of risk, as seen in **figures 1.2** and **1.3**. Symptoms of an inadequate or problematic risk culture are listed in **figure 1.4**.

Figure 1.2—Elements of Risk Culture

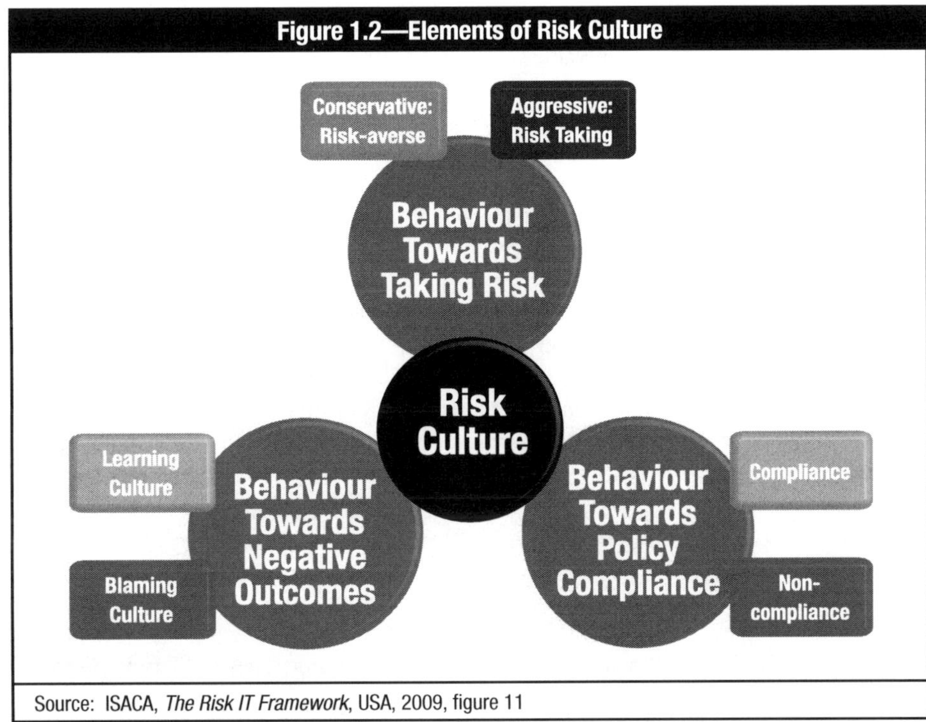

Source: ISACA, *The Risk IT Framework*, USA, 2009, figure 11

Figure 1.3—Description of Risk Culture Elements

Element	Description
Behavior toward taking risk	How much risk does the organization believe that it can absorb, and what specific risk is it willing to take?
Behavior toward policy compliance	To what extent do people within the organization embrace or comply with policy?
Behavior toward negative outcomes	How does the enterprise deal with losses, missed opportunities, and other negative outcomes? Does it learn from them and try to adjust, or does it assign blame without treating the root cause?

Figure 1.4—Symptoms of an Inadequate or Problematic Risk Culture

Symptom	Description
Misalignment between the risk appetites stated in policy and reflected in behavior	Management promulgates policies that mandate an approach to risk that is largely or completely different from demonstrable behavior by and within the organization.
Existence of a "blame culture"	In a blame culture, discussions focus on symptoms and accountability for problems. Root causes are rarely identified. Business units tend to blame IT when projects are not delivered on time or do not meet expectations. In doing so, they fail to realize how the business unit's involvement upfront affects project success. In extreme cases, the business unit may assign blame for a failure to meet the expectations that the unit never clearly communicated. The "blame game" only detracts from effective communication across units, further fueling delays. Executive leadership must identify and quickly control a blame culture if collaboration is to be fostered throughout the enterprise.

1.2.2 Risk Communication

The method and openness of risk communications play a key role in defining and understanding the risk culture of the organization. Communication is important because it removes the uncertainty and doubts concerning risk management. If risk is to be managed and mitigated, it must first be discussed and effectively communicated to the various stakeholders and personnel throughout the organization in ways that are appropriate for their respective roles. The benefits of open communication on risk include:

- More informed risk decisions by executive management due to an improved understanding of actual exposure and potential business impact
- Greater awareness among all stakeholders of the importance and value of integrating risk management into their daily duties
- Transparency to external stakeholders regarding both the actual levels of risk facing the organization and the risk management processes in use

The consequences of poor communication on risk include:

- A false sense of confidence at all levels of the enterprise and unintentional acceptance due to ignorance of risk that exceeds the organization's risk appetite
- Lack of direction or strategic planning to mandate risk management efforts
- Unbalanced communication of organizational risk to external stakeholders, potentially leading to incorrect and negative perceptions by third parties such as:
  - Clients
  - Investors
  - Regulators
- The perception that the organization is trying to hide risk from stakeholders

Figures 1.5 and **1.6** describe the broad array of information flows and major types of IT risk information that should be communicated.

Figure 1.5—Risk Communication Components

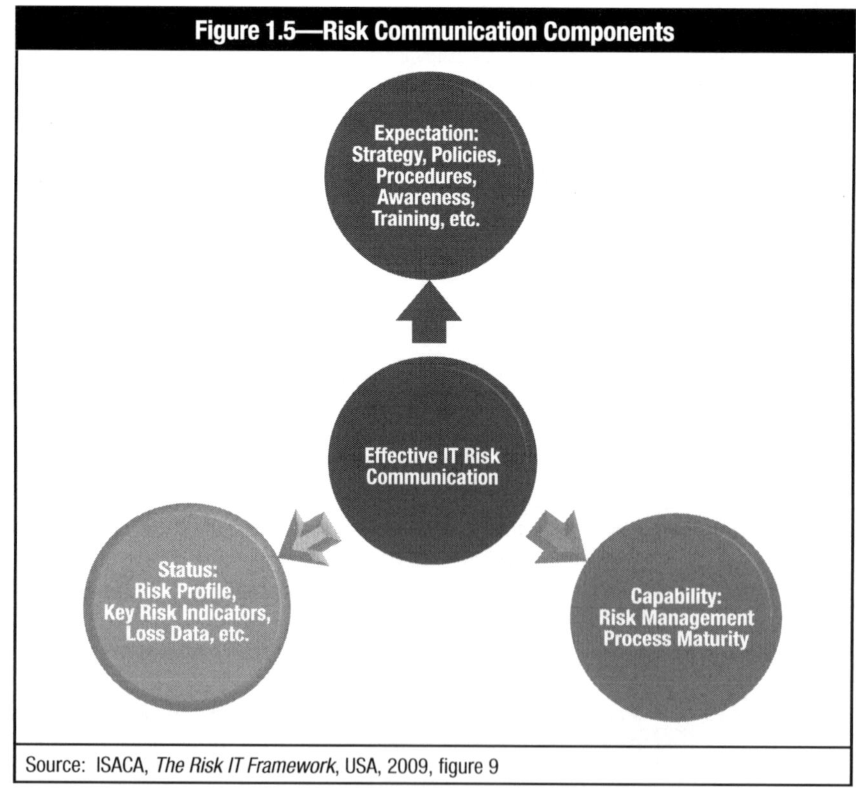

Source: ISACA, *The Risk IT Framework*, USA, 2009, figure 9

Figure 1.6—Risk Components to Be Communicated	
Type	**Description**
Expectations from risk management	Risk strategy, policies, procedures, awareness training and continuous reinforcement of principles. This is essential communication regarding the enterprise's overall strategy toward IT risk that: • Drives all subsequent efforts on risk management • Sets the overall expectations about the risk management program
Current risk management capability	• Allows for monitoring of the state of the "risk management engine" in the enterprise • Is a key indicator for good risk management • Has predictive value for how well the enterprise is managing risk and reducing exposure
Status	Includes the actual status with regard to IT risk, including information such as: • The risk profile of the enterprise (i.e., the overall portfolio of [identified] risk to which the enterprise is exposed) • Key risk indicators (KRIs) to support management reporting on risk • Event/loss data • The root cause of loss events • Options to mitigate risk (including cost and benefits)

The Value of Communication

The discipline of risk management is related to audit, business continuity management and security, and having a reporting relationship that facilitates communication regarding risk, incidents, vulnerabilities and assets among those groups can result in a greater level of accuracy for the risk management process.

Communication of risk incidents is an essential part of the risk management process. Not every risk incident needs to be reported, but a lack of communication can be a sign that an organization is not healthy or stable. If managers are not open and transparent with regard to issues and operational failures, decisions may be made on the basis of incomplete or inaccurate information. The risk practitioner should seek to develop lines of communication and reporting so that information is available to management on a timely basis and encourage communication of even negative activities when appropriate.

1.3 ELEMENTS OF RISK

Risk identification requires the documentation and analysis of the elements that comprise risk:

1. Consequences associated with specific assets
2. A threat to those assets, requiring both intent (motivation) and capability
3. Vulnerability specific to the threat

Each element of risk needs to be considered both individually and in aggregate. Identification of risk depends on successful identification of assets, threats to those assets and the vulnerabilities that the assets contain. Key terms for the identification of threats and vulnerabilities are listed in **figure 1.7**.

Figure 1.7—Risk Identification Terminology	
Term	**Definition**
Asset	Something of either tangible or intangible value that is worth protecting, including people, information, infrastructure, finances and reputation
Asset value	What the organization or another party would pay to take possession of an asset or deny access to it by others. Assets are typically valued on a monetary basis.
Impact	Magnitude of loss resulting from a threat exploiting a vulnerability
Impact analysis	A study to prioritize the criticality of information resources for the enterprise based on costs (or consequences) of adverse events. In an impact analysis, threats to assets are identified and potential business losses determined for different time periods. This assessment is used to justify the extent of safeguards that are required and recovery time frames. This analysis is the basis for establishing the recovery strategy.
Impact assessment	A review of the possible consequences of a risk
Likelihood	The probability of something happening

Figure 1.7—Risk Identification Terminology *(cont.)*	
Term	**Definition**
Threat	Anything (e.g., object, substance, human) that is capable of acting against an asset in a manner that can result in harm
Threat agent	Methods and things used to exploit a vulnerability, such as determination, capability, motive and resources
Threat analysis	An evaluation of the type, scope and nature of events or actions that can result in adverse consequences; identification of the threats that exist against enterprise assets
Threat vector	The path or route used by the adversary to gain access to the target
Vulnerability	A weakness in the design, implementation, operation or internal control of a process that could expose the system to adverse threats from threat events
Vulnerability analysis	A process of identifying and classifying vulnerabilities
Vulnerability scanning	An automated process to proactively identify security weaknesses in a network or individual system

1.3.1 Risk Factors

Risk is a combination of several factors that interact to cause damage to the assets of the organization, as shown in **figure 1.8**.

Figure 1.8—Risk Factors

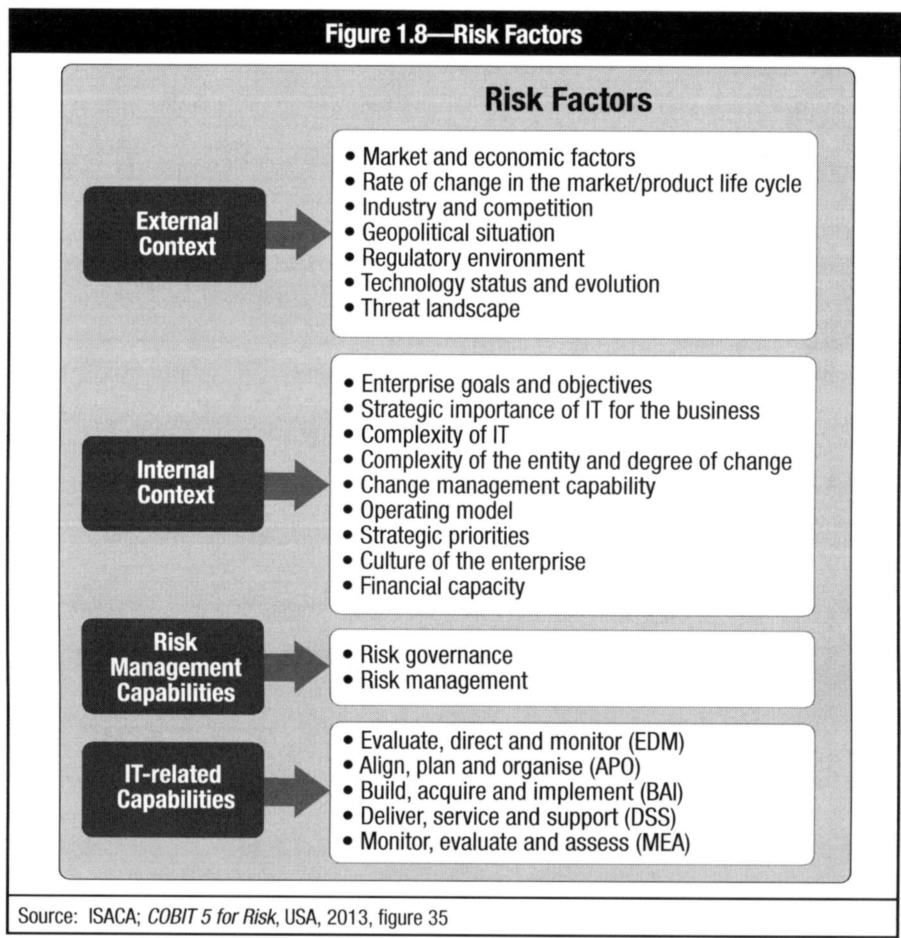

Source: ISACA; *COBIT 5 for Risk*, USA, 2013, figure 35

As shown in **figure 1.9**, threat agents use threats to attack assets via vulnerabilities. Threat agents that lack threats do not pose significant risk. Similarly, properly protected assets that are not vulnerable to threats being issued present no risk. For example, if a virus has been written to exploit a vulnerability on a system, the virus can only have an impact if the system has not been patched to eliminate the vulnerability. After the patch has been applied, the virus poses no threat to the system.

Figure 1.9—Risk Factors

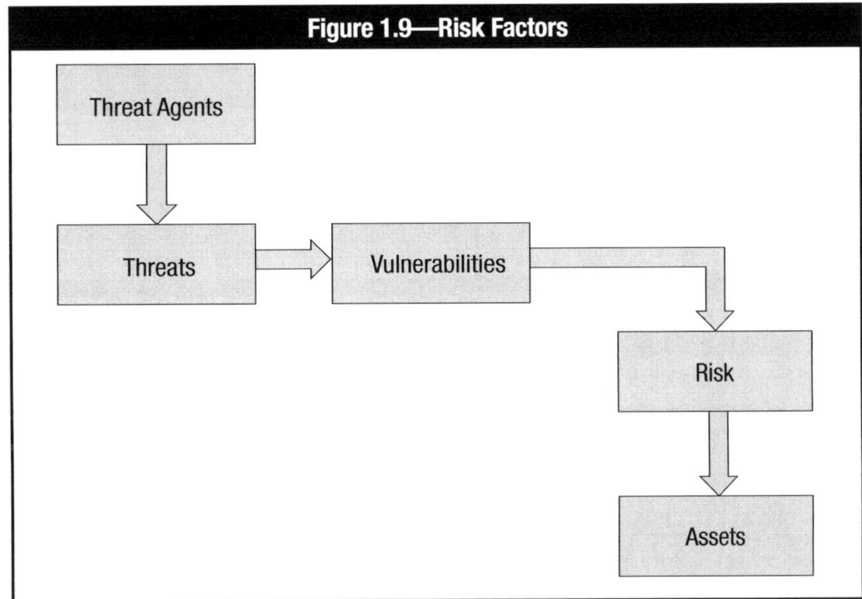

Knowledge of threats and the motivations, strategy and techniques of those who perpetrate threats is key, so that a threat may be managed before it becomes reality. An enterprise needs to know its own weaknesses, strengths, vulnerabilities and the gaps in the security fabric. Security must be woven into each business process, each IT system and all operational procedures. Any gaps in the security fabric can easily be exploited. The better understanding the risk practitioner has of the mind of the attacker or the source of the threat, the more effective the risk management activities will be in controlling the threat.

The threat and vulnerability landscape is always changing. People move, equipment wears out, controls weaken, new threats emerge and security awareness dulls through inaction. The risk practitioner must proactively seek out threats, vulnerabilities and controls through regular assessments, testing, observation and analysis in order to be able to identify risk. When evaluating risk, it is important to consider all types of risk, including technology-related risk. This type of risk can include a threat agent intercepting, interrupting, modifying or fabricating data on information assets, as well as the risk to the people and procedures that are used in acquiring, implementing, maintaining and disposing of that technology.

Risk evaluation is the measurement of risk, and the entire risk environment must be evaluated. The risk environment includes:
- The context, criticality and sensitivity of the system or process being reviewed
- The dependencies and requirements of the system or process being reviewed
- The operational procedures, configuration and management of the system or technology
- The training of the users and administrators
- The effectiveness of the controls and monitoring of the system or business process
- The manner in which data and system components are decommissioned

Risk is often influenced more by lack of training than by lack of equipment. In many cases, the risk is related to the way the equipment is operated more than whether the right tools are available.

1.3.2 Assets

An asset is something of either tangible or intangible value that is worth protecting. Examples of assets commonly found in organizations include:
- Information
- Reputation
- Brand

- Intellectual property
- Facilities
- Equipment
- Cash and investments
- Customer lists
- Research
- People
- Service/business process

People

Many organizations are vulnerable to the loss of a key employee who may be the only person with knowledge in a certain area or specific expertise. Failures by management to identify key employees and ensure that they are supported through cross-training and incentive programs are unfortunately common. Whether the loss occurs as a result of retirement, illness or recruitment by another organization, it may leave the organization in a precarious and vulnerable position.

Technology

Technology changes rapidly, and new technologies are always being developed. In addition to being aware of new technologies and the risk that they pose, the risk practitioner should also consider the risk associated with outdated technology, which is often overlooked. Equipment that is no longer supported or past its mean time between failure (MTBF) may be particularly vulnerable. Lack of patching and updating of systems and applications leaves them vulnerable to malware or misuse, and older systems may require expertise that is not readily available to maintain (potentially increasing key-person dependency). Older systems may also lack documentation, may be reliant on one vendor to maintain or be difficult to support in terms of replacement parts.

When systems are scheduled for disposal, they may contain sensitive data. The risk practitioner should be sure that procedures are in place to securely delete such data and that these procedures are followed. The method of secure disposal appropriate for a particular system is based partially on its physical form and partly on the sensitivity of the data on the device; common methods of destroying data include overwriting, degaussing and physical destruction of the equipment. If the data may be needed later, the organization may need to retain a copy of the software or even a legacy system to read them. In the case of encrypted data, keys or password files must also be stored. Failure to remove retired systems from backup schemes or business continuity or disaster recovery plans may also affect the integrity of those operations.

Data

Many organizations consider data to be extremely valuable. Customer lists, financial data, marketing plans, human resources (HR) data and research are some examples of data-related assets that are typically protected. Data may be sensitive, critical or both; sensitive data must be protected from disclosure or modification, while critical data must be protected from destruction or loss. The systems that host, process or transmit the data must ensure that data are protected at all times, in all forms (paper, magnetic storage, optical storage, reports, etc.) and in all locations (storage, networks, filing cabinets, archives, etc.).

Intellectual Property

Trademarks, copyrights, patents, trade secrets and other items associated with the reputation and goodwill of the organization, as well as research that leads to a new product, are examples of intellectual property. Intellectual property is a class of information that is treated with special care because it may represent the future earnings potential of the organization. Failure to protect intellectual property may result in the loss of competitive advantage. All employees and business partners should be bound by nondisclosure agreements (NDAs) and reminded of their responsibility to protect the intellectual property of the organization and handle it properly. This may include strict access controls, shredding of documents, caution when discussing information in a public location and encryption of data on portable media. Key terms related to intellectual property are described in **figure 1.10**.

Figure 1.10—Intellectual Property Terms	
Term	**Definition**
Trademark	A sound, color, logo, saying or other distinctive symbol that is closely associated with a certain product or company. Some trademarks are eligible for registration.
Copyright	Protection of any work that is captured in a tangible form (e.g., written works, recordings, images, software, music, sculpture, dance, etc.)
Patent	Protection of research and ideas that led to the development of a new, unique and useful product to prevent the unauthorized duplication of the patented item
Trade secret	A formula, process, design, practice or other form of secret business information that provides a competitive advantage to the organization that possesses the information

Business Processes

Inefficient or outdated business processes may pose a risk by making organizations uncompetitive. Business processes should be flexible enough to adapt to changes in the market or technology.

Asset Valuation

Not all assets are equally important; many are absolutely critical to business operations, while others are conveniences. The risk practitioner should make an effort to determine the importance of assets in the context of organizational activities so that priority may be given to protecting the most important assets first and addressing less significant assets as time and budget allow. Effective valuation also protects the organization from paying more in protection than the net worth of the asset.

The calculation of asset value is not as straightforward as it may initially appear. Many organizations use a quantitative approach that assigns a monetary value, which can be difficult when the value influences or drives intangibles such as confidence, morale or market perception. For example, if a product, company or region is associated with poor quality, environmental negligence or fraudulent activity, a high-quality product may be perceived as substandard due to this market perception. This negative perception may persist for many years before the company is able to recover.

Contributing factors to calculating asset value include:
- Financial penalties for legal noncompliance
- Impact on business processes
- Damage to reputation
- Additional costs for repair/replacement
- Effect on third parties and business partners
- Injury to staff or other personnel
- Violations of privacy
- Breach of contracts
- Loss of competitive advantage
- Legal costs

When calculating asset value, one technique is to base it on the impact of a loss of confidentiality, integrity and availability (CIA). This approach attempts to relate the impact to easily understood terminology. In order for this to work, the values for each level must be clearly stated and used in the same way by all departments.

1.3.3 Threats

Threats can be external or internal, intentional or unintentional. They may be caused by natural events or political, economic or competitive factors. Threats always exist and are typically beyond the direct control of the risk practitioner or asset owner. Not all conceivable threats need to be considered by every organization. For example, an organization that operates in a region with a seismic rating of zero does not have to document exposure to volcanoes or earthquakes. However, it is important to identify the various types of threats that do apply and may be used to compromise systems or otherwise affect the organization.

Threats may be divided into multiple categories, including:[5]

- Physical
- Natural events
- Loss of essential services
- Disturbance due to radiation
- Compromise of information
- Technical failures
- Unauthorized actions
- Compromise of functions

Several excellent lists of threats are available that can be used by a risk practitioner to ensure that all threats have been considered. These include NIST SP 800-30 Revision 1: *Guide to Conducting Risk Assessments, available at csrc.nist.gov*, and ISO/IEC 27005:2008 Appendix C, available at *iso.org*.

The risk practitioner should document all the threats that may apply to the systems and business processes under review. This requires using the resources noted previously, but should also include examining the cause of past failures, audit reports, media reports, information from national computer emergency response teams (CERTs), data from security vendors and communication with internal groups.

Threats may be the result of accidental actions, intentional/deliberate actions or natural events. Threats may originate from either internal or external sources, and actual attacks may leverage a combination of both internal and external sources. Where threats may be directed by individual agents, the risk practitioner should know and remain aware that such threat agents tend to be imaginative, creative and determined and will explore new methods and avenues of attack. To counter threat agents, the risk practitioner must also be determined and creative, and seek to discover as many threats as possible. An unidentified threat is one for which the organization is more likely to be unprepared and vulnerable than a threat that is well documented.

Sources for information regarding threats are listed in **figure 1.11.**

Figure 1.11—Sources for Threat Information	
Service providers	Insurance companies
Threat monitoring agencies	Product vendors
Security companies	Government publications
Audits	Assessments
Management	Users
Business continuity	Human resources
Finance	Media

Internal Threats

Although many organizations label their employees and other staff to be their best asset, employees may be unhappy if they are inadequately trained, treated poorly or not given enough time to do their jobs properly. Disillusionment and resentment can lead to a higher risk of errors, negligence and more conscious actions such as theft. Key personnel also may be drawn to another company and leave serious gaps in the knowledge and skills needed to operate systems effectively.

Employees are the cause of a significant number of business impacts, which can be intentional and unintentional. A disgruntled employee may intentionally compromise systems or release data that expose the organization to legal or reputational risk. Employees may be convinced, bribed or threatened to disclose trade secrets for ideological or economic reasons. Many employees have a level of access to systems and data that far exceeds their actual job requirements, which can be exploited in an attack. The solution to the employee problem, therefore, lies at least in part in the application of need-to-know and least privilege, but it is an imperfect solution. Any system has trusted insiders, and one of them choosing to violate trust is difficult to either predict or prevent.

The typical malicious insider is a current or former employee, contractor or other business partner who has or had authorized access to an organization's network, system or data and intentionally intercepted (exfiltrated), interrupted, modified or fabricated data on the organization's information systems. The first step in addressing personnel threats is to start with the hiring process and to review the qualifications and attitude of prospective employees. Employment candidates may have submitted incorrect information on job applications and claimed education, certification or experience that they did not actually possess. At the time of hiring, the employee should be required to sign a nondisclosure agreement and be advised of the ethics and policies of the organization, and a review of references and performance of background checks may be worthwhile where permitted by law.

Throughout employment, employees should be reminded of organizational policies and their responsibilities through awareness sessions and regular management reviews. One of the best employee-based controls is to interact with employees to understand any frustrations, complaints or issues that they may be facing and to seek to resolve those issues. During times of strike, layoffs, mergers, relocation and reorganization, an employee is more likely to be a risk. An employee who has been recently demoted or bypassed for a promotion is also a risk.

At the end of employment, an employee should return all organizational assets, including identification badges, equipment (e.g., laptops, mobile phones, access cards, etc.) and uniforms so that he/she cannot use those to gain unauthorized access in the future. In addition, systems, network and facility access should be removed immediately prior to the employee's departure to minimize the potential for crimes of opportunity.

External Threats

In a networked environment where data are stored offsite or hosted by cloud service providers, threats to information systems from outside of the organization can originate from anywhere and may take a number of forms, including, but not limited to, the following:

- Espionage
- Theft
- Sabotage
- Terrorism
- Criminal acts
- Software errors
- Hardware flaws
- Mechanical failures
- Lost assets
- Data corruption
- Facility flaws (freezing pipe/pipe burst)
- Fire
- Supply chain interruption
- Industrial accidents
- Disease (epidemic)
- Seismic activity
- Flooding
- Power surge/utility failure
- Severe storms

Natural events such as a flood, storm, earthquake or tornado are unpredictable and may be extremely damaging. The use of governmental data and weather monitoring services may identify the threats associated with natural events and allow the risk practitioner to take necessary steps to be prepared.

An external personnel threat includes a hacker, a thief or advanced persistent threat (APT) that is skilled and determined to break into systems for military or economic purposes. The term APT refers to advanced, highly skilled attackers that are determined (persistent) in their attempts to exploit systems and networks. The increased skills available to the hacking community and the effectiveness of the tools they possess make the risk of compromise much more significant. APTs may be sponsored by governments, organized crime or competitors.

Most breaches are the result of targets of opportunity, not determined attacks. As seen in the annual reports from Verizon and other organizations, many organizations are breached because they were discovered to be easy targets and threat agents took advantage of their vulnerabilities.

Emerging Threats

Indications of emerging threats may include unusual activity on a system, repeated alarms, slow system or network performance, or new or excessive activity in logs. In many cases, compromised organizations have evidence of emergent threats in their logs well in advance of the actual compromise, but the evidence is either not noticed or acted on. Lack of effective monitoring, when combined with a threat, is a combination that can lead to a breach.

Most technologies are built with an emphasis on function and purpose without due consideration for the security implications. As a result, new technology tends to be a source of new vulnerabilities and may even itself be a threat agent within an information system. The risk practitioner must be alert to the emergence of new technologies and prepare for their introduction into the organization, particularly if these technologies promise cost savings or competitive advantage. Bring your own device (BYOD) is an example of a revolution in how organizations view technology assets, one whose risk is self-evident but that has tempted a wide variety of organizations by promising to greatly reduce the cost of both initial procurement of IT assets and the rate at which they need to be refreshed. A threat strategy that emphasizes rejection of new technology is unlikely to remain in place long beyond the point at which something gains executive sponsorship.

1.3.4 Vulnerabilities

Vulnerabilities are weaknesses, gaps or holes in security that provide an opportunity for a threat or create consequences that may impact the organization.

The *National Institute of Standards and Technology (NIST) Special Publication 800-30 Revision 1: Guide to Conducting Risk Assessments* provides a list of vulnerabilities to consider as well as "predisposing conditions" that may lead to the rapid or unpredictable emergence of new vulnerabilities. Many vulnerabilities are conditions that exist in systems and must be identified so that they can be addressed. The purpose of vulnerability identification is to find the problems before they are found by an adversary and exploited, which is why an organization should conduct regular vulnerability assessments and penetration tests to identify, validate and classify its vulnerabilities. Where vulnerabilities exist, there is a potential for risk.

Note: Other sources of vulnerabilities include:
- National Vulnerability Database at *nvd.nist.gov*
- Common Weakness Enumeration at *nvd.nist.gov/cwe.cfm*
- Harmonized Threat and Risk Assessment at *www.cse-cst.gc.ca*
- Contingency Planning and Management at *contingencyplanning.com*
- Open Web Application Security Project (OWASP) at *www.owasp.org*

Network Vulnerabilities

Network vulnerabilities are often related to misconfiguration of equipment, poor architecture or traffic interception. Misconfiguration is a common problem with network equipment that is not properly installed, operated or maintained. Any open services are a potential attack vector that can be exploited by an attacker, so network equipment should be hardened by disabling any unneeded services, ports or protocols.

Traditionally, networked systems have been limited to devices whose primary purpose was to provide computing power, either for direct processing of data or to regulate the flow of data within the network. However, as the cost of networking technology drops in terms of dollars, power and physical size, the ability to transmit and receive data is being incorporated into a growing number of devices whose primary purposes are not of a data processing or regulatory nature, such as home appliances and automobiles. These devices create new vulnerabilities for the network. The resulting Internet of Things is emerging as both a tremendous opportunity for increasing convenience and a vast new target for threat agents. Therefore, the risk practitioner should ensure he/she is knowledgeable on emerging technologies and their implications for future use.

Physical Access

The *Verizon 2015 Data Breach Investigations Report*[6] estimates that approximately 15 percent of all incidents relate to physical theft or loss of equipment or data. However, physical security goes far beyond theft. Threat agents who are able to gain physical access to systems have the potential to bypass nearly every other type of control. With access to server rooms, network cabling, information systems equipment and buildings, an attacker can circumvent passwords, install "skimmers" to intercept data communications and take logical ownership of systems or devices. With more than 20 percent of incidents related to misuse of systems by insiders who may already have an advantage in terms of gaining physical access, the need for strong physical security is clear and growing.

Testing for physical security vulnerabilities includes testing locks, security guards, fire suppression systems, heating ventilation and air conditioning controls, lighting, cameras, and motion sensors.

Applications and Web-facing Services

Applications in general, web applications in particular, are among the most common entry points currently used by attackers. Many applications are written to support business function without due regard to the security requirements of the applications and may be vulnerable to buffer overflows, logic flaws, injection attacks, bugs, incorrect control over user access and many other common vulnerabilities. The risk practitioner can use tools from the Open Web Application Security Project (OWASP)[7] to test web-facing applications for well-known and documented vulnerabilities whose presence can lead to data compromise or system failure.

Applications may also be vulnerable on account of poor architecture. Applications that process payment card data or other sensitive information or store those data in an insecure location, such as in a demilitarized zone (DMZ), may come under repeat or even continuous attack, increasing the likelihood that an attacker will find a vulnerability to exploit. The risk practitioner should understand the network architecture and question decisions that appear to create unnecessary exposure.

Utilities

Information systems rely on controlled environmental conditions, including clean and steady power and controls over humidity and temperature. The risk practitioner should ensure that the organization is prepared for power failure or other environmental conditions in order to avert conditions that may lead to system failure. Having an uninterruptible power supply (UPS), backup generators and surge protectors can protect equipment from damage or failure, but these systems require more than an initial investment. For instance, a UPS must be validated to ensure it has adequate power to run critical systems until a backup generator can come online and then tested periodically, while generators must be checked for fuel as well as mechanical function.

Supply Chain

Many organizations rely on products, raw materials and supplies that originate from various locations around the world, and any interruption in the supply chain may affect their ability to function. For example, a shortage in supply of fuel to an airport would impact all flight operations. Reliance upon a supply chain should be documented even if it is understood that the risk is one that is accepted by senior management.

Processes
Maintaining operational integrity often requires procedures spanning a variety of topics, including incident management, identity management, change control, patch management and project planning. These processes must be defined and deployed in a consistent manner across the organization. Without them, or if they are deployed haphazardly, the organization may be at risk of inconsistent management and results, lack of governance and reporting, and failure to ensure compliance with regulations.

Equipment
As equipment ages, it becomes less efficient, effective, and able to support business functions. Equipment is often provided at the time of production with MTBF rating that indicates its anticipated life span and when it should be scheduled for removal or replacement.

Cloud Computing
Outsourcing of application hosting and data processing has been growing in popularity for decades. However, the growing availability and plunging cost in recent years of both high-speed connectivity and servers capable of hosting virtual machines has transformed the practice. Services hosted in third-party data centers now have astonishing redundancy, automatic failover and service level agreements (SLAs) that guarantee uptime in excess of 99.9 percent—all through an on-demand pricing structure that can undercut in-house operations by an order of magnitude. These characteristics make the use of cloud-based services very attractive to many organizations. There are several cloud-based service models, as seen in **figure 1.12**.

Figure 1.12—Cloud Deployment Models

Private cloud	• Operated solely for an enterprise • May be managed by the enterprise or a third party • May exist on- or off-premise
Public cloud	• Made available to the general public or a large industry group • Owned by an organization selling cloud services
Community cloud	• Shared by several enterprises • Supports a specific community that has a shared mission or interest • May be managed by the enterprises or a third party • May reside on- or off-premise
Hybrid cloud	• A composition of two or more clouds (private, community or public) that remain unique entities, but are bound together by standardized or proprietary technology that enables data and application portability (e.g., cloud bursting for load balancing between clouds)

Source: ISACA, *IT Control Objectives for Cloud Computing: Controls and Assurance in the Cloud*, USA, 2011, figure 1.3

Cloud computing offers many business advantages, but the risk practitioner should be conscious that the outsourcing of data processing does not remove the liability of the outsourcing organization to ensure proper data protection. Also, while larger data centers that are established to serve hundreds or thousands of customers are typically better protected than stand-alone operations, the ability of the outsourcing organization to see what goes on inside the company that hosts its data center may be very limited. When one considers that the cloud data centers are large, visible and tempting targets for the best-funded and most experienced threat agents, it is clear that cloud computing represents risk as well as opportunity. The decisions of whether to leverage cloud services, which systems should be transferred and which might be better retained within the control of the organization should be made by senior management. In order to assist them in making this decision, the risk practitioner should provide senior management with the necessary information so they can have a thorough understanding of the risk involved.

Big Data
Advances in the capability to perform analysis of data from various sources of structured and unstructured data allow enterprises to make better business decisions and increase competitive advantage. This change in analytics capabilities dealing with "big data" can introduce technical and operational risk, and organizations should understand that risk can be incurred either through adoption or non-adoption of these capabilities.

The risk of adopting or not adopting big data analytics is described by ISACA the following way:[8]

> *Technical and operational risk should consider that certain data elements may be governed by regulatory or contractual requirements and that data elements may need to be centralized in one place (or at least be accessible centrally) so that the data can be analyzed. In some cases, this centralization can compound technical risk.*
>
> *For example:*
> - ***Amplified technical impact**—If an unauthorized user were to gain access to centralized repositories, it puts the entirety of those data in jeopardy rather than a subset of the data.*
> - ***Privacy (data collection)**—Analytics techniques can impact privacy; for example, individuals whose data are being analyzed may feel that revealed information about them is overly intrusive.*
> - ***Privacy (re-identification)**—Likewise, when data are aggregated, semianonymous information or information that is not individually identifiable information might become non-anonymous or identifiable in the process.*

1.3.5 Vulnerability Assessment and Penetration Testing

A vulnerability assessment is a careful examination of a target environment to discover any potential points of compromise or weakness. Vulnerabilities that may be identified by an assessment include:
- Network vulnerabilities
- Poor physical access controls (e.g., buildings, offices)
- Insecure applications
- Poorly designed or implemented web-facing services
- Disruption to utilities (e.g., power, telecommunications)
- Unreliable supply chain
- Untrained personnel (HR)
- Inefficient processes (e.g., change control, incident handling)
- Poorly maintained or old equipment

Vulnerability assessment may be a manual process or use automated tools. Automated tools have the ability to filter large amounts of data and can be used to examine logs, compile and analyze data from multiple sources (e.g., a security incident and event management [SIEM] tool), examine the functions of a program and run test files or data against a tool such as a firewall or application. Where content is not easily quantified and requires judgment—including reviews of business processes, physical security and source code—a manual test is better (although automated tools for reviewing code continue to improve).

Additional information on penetration testing can be found in chapter 2, section 2.3.15 and chapter 4, section 4.5.3.

Vulnerability assessments may contain data that is not accurate, particularly false positives that indicate a vulnerability where none exists. To validate the results of a vulnerability assessment, the organization may conduct a penetration test against a potential vulnerability or attack vector. Penetration testing uses the same sort of tools as those used by a real adversary, which can help to establish the extent to which an identified vulnerability is a true weakness. Penetration tests may be conducted by internal or external teams, and rules can range from full knowledge of the infrastructure to a zero-knowledge tests in which the testing team has no knowledge of the infrastructure being attacked. An expert penetration testing team will be creative and attempt several types of tests to ensure that as many attack vectors as possible have been tested, generating a report that the risk practitioner can use in the process of risk identification.

1.3.6 Likelihood/Probability

The risk faced by organizations consists of some combination of known and unknown threats, directed against systems that have some combination of known and unknown vulnerabilities. A vulnerability assessment that identifies no vulnerabilities is not equal to the system being invulnerable in the absolute sense. It means only that the types of vulnerabilities that the assessment was intended to detect were not detected. Failure to detect a vulnerability may be the result of its absence, or it may be a false negative arising from misconfiguration of a tool or improper performance of a manual review. In the case of a zero-knowledge penetration test that fails to identify opportunities

to exploit a system, it may be that the team was unlucky or lacked imagination, which may not be true of an outside attacker.

Even if no known vulnerabilities exist within the system, the system may still remain vulnerable to unknown vulnerabilities, more of which are discovered every day (and some of these are stored or sold for future use as "zero-day" exploits). The likelihood of an attack is often a component of external factors, such as the motivation of the attacker, as shown in **figure 1.13**.

Figure 1.13—Influencing Risk Factors

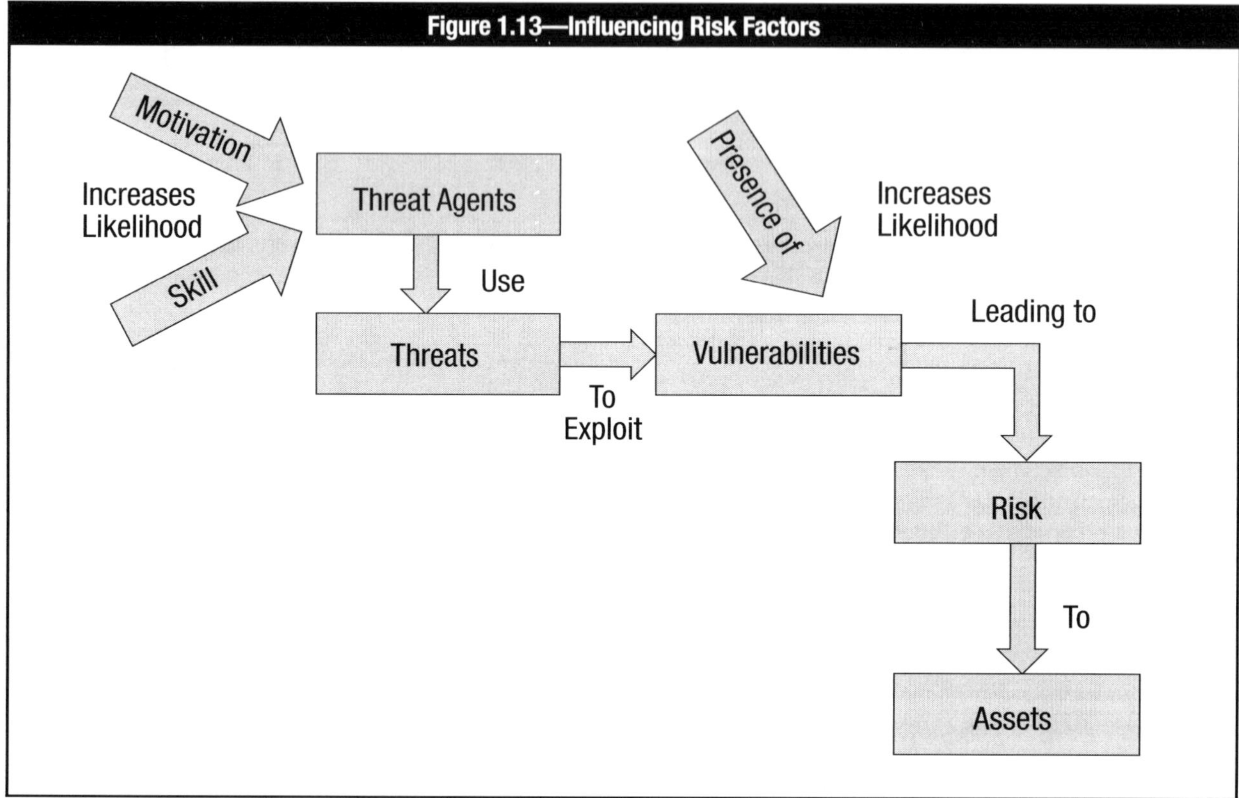

Given the combination of unknown threat and unknown vulnerability, it is difficult for the risk practitioner to provide a comprehensive estimate of the likelihood of a successful attack. Vulnerability assessments and penetration tests provide the risk practitioner with valuable information on which to partially estimate this likelihood, because:

- Although the presence of a vulnerability does not guarantee a corresponding threat, the all-hours nature of information systems and the rapid speed of processing makes it much more likely that an information system will come under an assortment of attacks in a short time than would be true of a physical system
- A vulnerability known to an assessment tool is also knowable to threat agents, all but the most elite of whom tend to build their attacks to target common vulnerabilities.
- The presence of one or more known vulnerabilities—unless these have been previously identified and the risk is accepted by the organization for good reason—suggests a weakness in the overall security program.

The risk practitioner should ensure that senior management does not develop a false sense of security as a result of vulnerability assessments and penetration tests that fail to find vulnerabilities, but both forms of testing do provide insight into the organization and its security posture.

1.4 INFORMATION SECURITY RISK CONCEPTS AND PRINCIPLES

IT risk is often linked to information security, which is the protecting of information and information systems (including technology) from risk events. Information security controls are based on risk, and risk is the primary justification used to support information security activities.

Some organizations have a mature risk management process that is able to quantify risk with a high level of accuracy. In other organizations, the term "risk" may be poorly understood and represent an emotional value based on perception. Risk is often difficult to measure, and because it is based on concepts such as likelihood and impact, it can be hard to quantify. In most cases, the predicted impact should an undesired event occur results in only direct, immediate effects, whereas actual events may have repercussions that affect tangential aspects of the business or are discovered long after the event has occurred.

Likelihood (also called probability) is the measure of frequency of which an event may occur, which depends on whether there is a potential source for the event (threat) and the extent to which the particular type of event can affect its target (vulnerability), taking into account any controls or countermeasures that the organization has put in place to reduce its vulnerability. In the context of risk identification, likelihood is used to calculate the risk that an organization faces based on the number of events that may occur within a given time period (often an annual basis). Specific factors that can affect likelihood include:

- **Volatility**—Unpredictability, also referred to as dynamic range; the degree to which conditions vary from one moment to another, making projections difficult
- **Velocity**—Speed of onset, a measure of how much prior warning and preparation time an organization may have between the event's occurrence and impact, which itself can be split into speed of reaction and speed of recovery
- **Proximity**—The time from the event occurring and the impact on the organization
- **Interdependency**—The degree to which materialization of two or more types of risk might impact the organization differently, depending on whether the events occur simultaneously or consecutively
- **Motivation**—In cases involving an active/sentient threat, the extent to which the perpetrator of the threat wants to succeed, which may result in a higher chance of success
- **Skill**—The ability brought to bear by the perpetrator of an active/sentient threat relative to other perpetrators
- **Visibility**—The extent to which a vulnerability is known, which can make it a more likely target of attack

In addition to likelihood, there are two forms of impact that the risk practitioner should consider:

1. Impact due to the loss or compromise of information
2. Impact due to the loss or compromise of an information system

The risk practitioner may find it helpful to evaluate an information system or data source on the basis of confidentiality, integrity and availability (CIA)—the three principles from which all other characteristics of information security are derived—to identify whether a compromise or failure in one or more of these areas would affect the organization. For purposes of risk identification, simply noting that an impact may occur is sufficient; during risk assessment, the practitioner may draw on these findings to develop a more detailed analysis that includes the particular extent of impact from each area of compromise.

It is important for the risk practitioner to have a solid understanding of what is meant by CIA, particularly the interrelationship between the three principles and a fourth—nonrepudiation—that is derived from them. There are typically trade-offs between the three principles in any system such that one cannot simultaneously increase one of them without either decreasing at least one of the others or substantially increasing cost; for instance, increasing confidentiality typically increases processing time, which reduces availability. The CIA triad shown in **figure 1.14** illustrates this relationship.

Figure 1.14—The CIA Triad

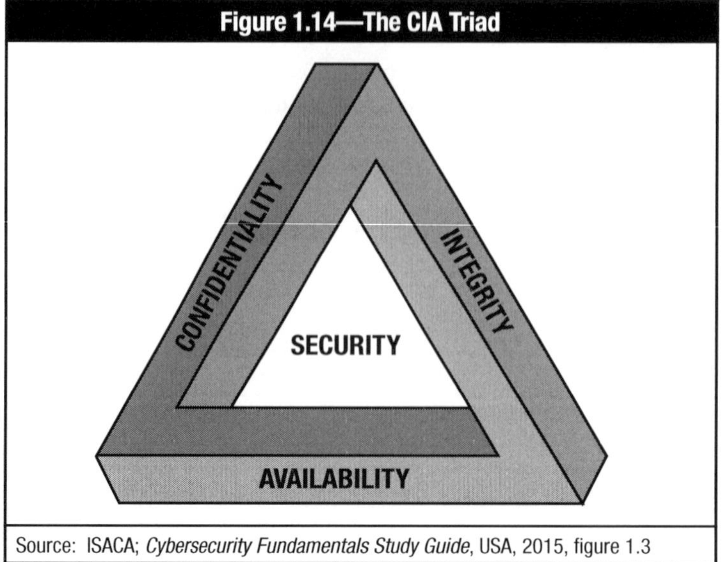

Source: ISACA; *Cybersecurity Fundamentals Study Guide*, USA, 2015, figure 1.3

Confidentiality, integrity, availability, and nonrepudiation are briefly discussed in the sections that follow, and are covered in detail in chapter 3 Risk Response and Mitigation.

1.4.1 Confidentiality

Confidentiality refers to the secrecy and privacy of data. A breach of confidentiality means the improper disclosure of information, such as disclosing information to an internal or external resource that was not authorized to access the information. Confidentiality is often discussed in terms of the need to protect classified military secrets or other categories of sensitive information, such as personally identifiable information (PII), personal health information (PHI) or intellectual property, and specific types of data protected by law. However, the basic principle of confidentiality is disclosure to any recipient not authorized access by the information owner.

In identifying risk related to confidentiality, the risk practitioner should look for policies or behaviors that violate either or both of the following principles:

1. **Need to know.** Individuals should be given access only to information that is needed in order for them to perform their job functions. For example, most users do not need to see credit card numbers; therefore, the numbers are masked or hidden except for the last few digits, which protects the cardholder data from being compromised by a person that did not need to see the entire credit card number.
2. **Least privilege.** The level of data access afforded individuals or processes should be the minimum needed to perform their job functions. For instance, someone who needs to review data should not be able to edit or delete it.

Policies, practices or behavior that does not reflect these principles should be a warning flag to the risk practitioner of possible risk related to confidentiality.

1.4.2 Integrity

Integrity refers to guarding against improper modification, exclusion or destruction of information, which requires the protection of information from improper modification by unauthorized users, authorized users and processes or activities operating on the system. Whenever data are changed in a manner other than that intended by the data owner, integrity is compromised.

Maintaining integrity is typically a rigorous process that relies upon multiple levels of error checking and verification that may be difficult for the risk practitioner to fully evaluate without technical expertise. The principle of least privilege is also critical when identifying risk related to integrity, particularly when it comes to access by system processed and other memory-resident applications. Whenever a user or process has the power to change or delete data, the ways in which it interacts with data must be carefully considered to determine whether there is an integrity risk.

1.4.3 Availability

Availability refers to providing timely and reliable access to information. Timely access to data is often important to business processes. In some cases, such as industrial control systems operating machinery or regulating power generation, near-real-time availability may be needed for safety and system operations.

When asking the business what levels of availability they require, the answer is often that the system and data are essential and require 100 percent availability. In most cases, this is not accurate, which requires the risk practitioner to work with the business to determine what the true level of availability requirements are and what the organization is willing to pay to have higher availability. One way to identify potential availability risk is to compare current levels of availability with required levels; where there is a gap, there is a risk.

1.4.4 Nonrepudiation

In information systems, activity is often either anonymous or only weakly traceable to a particular individual or process. Nonrepudiation refers to a positive guarantee that a given action was carried out by a given individual or process and is an important part of tracing responsibility and enforcing accountability. Digital signatures and certificate-based authentication in a public key infrastructure (PKI) are examples of controls whose proper implementation and use can provide nonrepudiation, while shared or generic logon credentials do not even provide the relatively weak assurance afforded by individual user names and passwords. For purposes of risk identification, the risk practitioner should look for evidence of nonrepudiation in situations in which actions may have significant impact on an organization, such as approval of production code, deletion of records or disbursement of funds.

1.5 THE IT RISK STRATEGY OF THE BUSINESS

IT risk is the business risk associated with the use, ownership, operation, involvement, influence and adoption of IT within an enterprise. Understanding the business's overall risk strategy helps the risk practitioner to guide development of an IT risk strategy that aligns with organizational goals and priorities. All IT risk must be measured not only by its impact on IT services but also by the impact of risk on business operations.

Many organizations have some form of risk management in place, but levels of formal definition, documentation and maturity can vary widely between organizations. One role of the risk practitioner is to investigate the business's current risk environment and confirm that documentation of controls, risk, audit and regulations that impact the determination of the level of risk is either available or that plans are underway to document these important aspects of the risk management program. Where plans do not already exist, it may fall to the risk practitioner to devise a strategy for creating the necessary documentation.

1.5.1 Types of IT-related Business Risk

Identifying risk requires a combination of imagination and analysis. The risk practitioner is most likely to identify a risk when it takes the form of something that he or she can already envision. For this reason, it is beneficial for the risk practitioner to know the types of business-related IT risk shown in **figure 1.15**.

The list of risk types shown in **figure 1.15** is not exhaustive. The risk practitioner should be committed to continuing education as a professional philosophy and be attentive to the concerns of business-process owners, who may conceive of potential problems specific to their particular areas of expertise that may not be immediately apparent to a generalist. Effective risk management is a collaborative process.

In addition, the risk practitioner should be aware that IT risk is only one component of the overall risk universe of an organization, which also includes risk associated with credit, regulatory compliance, environmental and labor considerations, and other factors substantially independent of IT. The need to consider the full spectrum of organization risk can overwhelm senior managers when individual risk is presented without adequate context. The risk practitioner should always recognize that what is critical to IT may or may not be the most critical consideration of the organization in the context of overall risk.

Figure 1.15—Business-related IT Risk Types	
Type	**Description**
Access risk	The risk that information may be divulged or made available to recipients without authorized access by the information owner, reflecting a loss of confidentiality
Availability risk	The risk that service may be lost or data are not accessible when needed
Infrastructure risk	The risk that the IT infrastructure and systems may be unable to effectively support the current and future needs of the business in an efficient, cost-effective and well-controlled fashion (includes hardware, networks, software, people and processes)
Integrity risk	The risk that data may be unreliable due to incompleteness or inaccuracy
Investment or expense risk	The risk that the IT investment fails to provide value commensurate with its cost or is otherwise excessive or wasteful, including the overall IT investment portfolio
Project ownership risk	The risk of IT projects failing to meet objectives through lack of accountability and commitment
Relevance risk	The risk that the right information may not get to the right recipients at the right time to allow the right action to be taken or the right decisions to be made
Schedule risk	The risk of IT projects taking longer than expected

1.5.2 Senior Management Support

Senior management support is tremendously important throughout the risk management process. With it, the risk management process is much more likely to have the budget, authority, access to personnel and information, and legitimacy that will provide a successful result. Without it, risk management is almost always unsuccessful. Senior management support should be visible and active, and executives should be willing to intervene when necessary to communicate the importance of risk identification efforts and the need for everyone to actively contribute to the success of the program.

1.5.3 Alignment With Business Goals and Objectives

Risk management depends on business goals and objectives. The risk practitioner must be careful not to consider risk from the perspective of one department or one process without also considering the risk to other departments, business partners or the overall business goals. It can be easy for IT risk practitioners to calculate only risk reflecting the impact on IT and forget to include the business processes that the IT systems support. The risk management program serves the organization, so the risk practitioner must remember to look beyond his/her own department.

Risk practitioners should keep in mind that some lines of business that are currently prominent may not be a part of the future strategy of the organization. The senior management team may already be planning to move away from one product line and toward a new emerging product. Objective commitment to the goals of the organization creates more long-term value than attempting to predict which business units may be prominent in the future.

The best way to understand the goals and objectives of the organization is to communicate with senior management. The risk practitioner should consult with senior management to gain an understanding of their vision and strategy regarding lines of business, new technologies, future growth or changes in priorities, keeping in mind that some information may be censored due to ongoing negotiations or confidential strategic plans. Risk considers both the potential gains and potential losses of pursuing new ventures, which may include financial loss, reputational damage, changes in employment agreements and entry into different regulatory environments. Knowing what executives are anticipating gives the risk practitioner insight into the potential risk universe of the organization as it evolves.

The underlying importance of risk management in relation to business goals and strategy is to ensure that the risk is closely aligned with, and integrated into, the strategy, vision and direction of the organization. Because risk management serves the organization, executives will ultimately choose a path that seems to offer the best prospects

for value creation and may do so despite the resistance of those involved in risk management. To avoid being seen as obstructionist, the risk practitioner should seek to:
- Understand the business
- Listen to the strategy
- Proactively seek out ways to secure new technologies and business processes
- Build relationships and communication infrastructure to weave risk management into each business process and new project
- Be aware of and mitigate the risk of change
- Work to create a culture that encourages integration of risk management into business processes
- Identify realistic ways that would make "yes" align with the risk appetite instead of saying "no"

History is a critical component of risk management. The events of the past must be considered in the light of the risk management effort, because an organization that has not taken steps to mitigate the future impact of events similar to those that have already occurred may be perceived by stakeholders as irresponsible. However, a narrow focus on the past may result in a lack of watching for new threats and future issues, with the organization falling into the habit of only fixing events once they have occurred. Alternately, an organization that was fortunate in the past may suffer from overconfidence, which can be dangerous and cause the organization to overlook serious problems that should be thoroughly addressed and mitigated. In this case, the preexisting conditions that led to the event may remain, setting the stage for a serious incident.

When an organization goes through a merger or acquisition, new risk emerges. Mergers create uncertainty and stress that can result in poor judgment or inappropriate actions by people worried about protecting their jobs, being moved into other positions, or having to compete for a new assignment. In many cases, mergers result in job loss, and some studies have suggested that employees who are about to lose their jobs are likely to take corporate assets and information with them. When employees have assets of the organization in their possession, including access to sensitive or protected information, the risk practitioner should actively increase the controls over sensitive information, be increasingly diligent to enforce monitoring and work with HR to identify potentially risk-laden scenarios.

1.5.4 Organizational Structures and Impact on Risk

The effectiveness of the risk management effort is often influenced by the positioning of the risk management function within the organizational structure. Ideally, risk management should be a function with enterprise scope, able to reach into all the parts of the organization and provide leadership, advice and direction. An effective risk management program provides a consistent way to manage risk and creates a risk framework that serves as a foundation for risk management for all departments and business functions. The organization should have established three lines of defense, with risk being managed by the front line and guided, directed, influenced and/or assessed by the second line and with independent oversight, review and monitoring by the third line. However, this is not always the case. More often than not, risk management is only one of the job responsibilities of a manager or is a small function buried deep within one area of the business.

Risk management may be applied to an entire organization as one or more formal risk management teams, or it may be practiced separately in each level of the organization or in regard to specific functions, projects and activities. Functional-level integration is valuable in the context of implementation, but decentralized models can make it unclear to whom the risk practitioner should report within the organization. Many companies do not have a dedicated chief risk officer (CRO), and of those that do, not all CROs report to the chief executive officer (CEO) or board of directors.

The size and diversity of the organization is a key factor in managing risk. Even in a small company operating in one primary market, where the management of risk is fairly straightforward, cultural differences between departments can be extreme. For example, the approach to risk in the finance department may be diametrically opposed to the approach to risk in the sales department. This requires the risk practitioner to work with the management team to develop an understanding of risk and a standard approach to measuring and mitigating risk.

In a large organization, risk management may be too large of a task for one department or team. If the organization operates in several countries or supports a wide variety of products or services, the risk management effort may need to be arranged by department, products, services or geographic region. In most cases, the organization of the risk management group should follow the same model and the organization of the business continuity management team.

RACI (Responsible, Accountable, Consulted, Informed)

There are four main types of roles that are involved in the risk management process:

- The individuals responsible for managing the risk
- The individuals accountable for the risk management effort
- The individuals who are consulted and provide support and assistance to the risk management effort
- The individuals who are informed of the risk management effort but may not necessarily be involved in its execution

The use of a RACI model can assist in outlining the roles and responsibilities of the various stakeholders. The purpose of a RACI model is to clearly show the relationships between the various stakeholders, the interaction between the stakeholders and the roles that each stakeholder plays in the successful completion of the risk management effort. **Figure 1.16** describes the components of a RACI model. An example of a RACI chart is shown in **figure 1.17**.

Figure 1.16—RACI Model

Assigned Role and Responsibility	Description
Responsible	Individuals tasked with getting the job done, performing the actual work effort to meet stated objectives.
Accountable	The single person liable or answerable for the completion of the task, who oversees and manages the person(s) responsible for performing the work effort, who may also play a role in the project. Accountability for a particular task should be assigned to a specific person in order to be effective.
Consulted	Individuals who provide input data, advice, feedback or approvals. Consulted personnel may be from other departments, from all layers of the organization, from external sources or from regulators.
Informed	Individuals who are informed of the status, achievement and/or deliverables of the task but who are often not directly responsible for the work effort.

Figure 1.17—Sample RACI Chart

Task	Senior Management	Steering Committee (Chair)	Department Managers	Risk Practitioner
Collect risk data	I	A	C	R
Deliver the risk report	I	A	I	R
Prioritize risk response	A	I	R	C
Monitor risk	I	A	R	C

1.5.5 Organizational Culture, Ethics and Behavior and the Impact on Risk

One of the first challenges faced by a risk practitioner is to determine the risk appetite of senior management. Examples of risk that management must evaluate and set criteria for accepting include deciding whether the organization should:

- Invest
- Take on a new line of business
- Develop a new product
- Open a new office
- Hire a new employee
- Invest in new hardware or software
- Upgrade existing applications
- Implement new controls

The risk practitioner should be aware that the risk appetite may change over time or vary based on the type of risk. Thus, the risk appetite requires periodic re-determination. Depending on market conditions, confidence, past successes or failures, global economics, reports in the media, availability of resources, new regulations or long-term strategy, the extent to which risk appetite changes may be dramatic. Culture and ethics also play a role in determining variances in risk appetite.

Culture

A culture drives the behaviors of personnel, and people will often act according to their environment. Risk culture is defined as the set of shared values and beliefs that governs attitudes toward risk taking, care and integrity and determines how openly risk and losses are reported and discussed. All employees of the organization should be aware of the risk culture, because knowing the risk culture provides a useful context for their own decisions and actions.

A person will often act according to beliefs, so creating an environment that addresses people's belief systems is often an effective method of changing behaviors as well. For instance, a culture of honesty and openness may reduce the risk of theft, inappropriate actions or attacks.

Even in small organizations, the risk practitioner should be aware that departments may have their own subcultures that differ from the organizational culture. Adherence to a subculture may lead to people taking more risk than management actually intends to accept; it may also manifest in people refusing to take risk when management would like them to do so. Subcultures may evolve out of industry norms, but they can also arise organically as a reaction to demonstrated actions. For instance, in an organization in which senior executives encourage employees to take risk but blame them when something is wrong, creativity and innovation may be less pronounced than would otherwise be the case.

Ethics

Risk is often impacted by professional ethics. It is easy to understand that an organization with poor ethical standards may be more susceptible to fraud or theft, but ethical risk may also apply to an organization that has poor management processes in place to identify errors, misuse or fraud. Ethics are related to an individual's perception of right and wrong and are not necessarily linked to the law. For example, in many organizations, it is acceptable to receive gifts from clients or suppliers, whereas in others it is not acceptable. To address the risk of a person violating the ethics policy of the organization, senior management must communicate policy to everyone and ensure that it is both visibly enforced and equally applicable to people at all levels of authority. Ethics policies that are unenforced or applied selectively are ineffective in directing behavior.

Ethics also applies to how people believe that they have been treated. A well-treated employee may be an active supporter of proper behavior. In contrast, an employee who feels that he or she has been poorly treated may seek revenge, with serious consequences.

1.5.6 Laws, Regulations, Standards and Compliance

Organizations are required to comply with the laws and regulations of the jurisdictions in which they operate and face penalties for failing to do so. It is important to know what laws apply to the organization and to understand their requirements, which can be challenging because many laws are open to interpretation and required levels of compliance are not always stipulated. For example, a law may require adequate protection of sensitive data without specifying what constitutes an adequate level of protection. Laws and regulations may also contradict one another, and an organization that operates in various regions of the world may find it difficult to comply with the all of the laws in each location.

Organizations that operate globally or even within different regions of one country may build a global program of policies and a control suite to handle the common regulations and then have a regional or nation-specific addendum to handle the exceptions and their controls. For instance, the European Union Data Protection Regulation 45/2001[9] requires:

1. Personal data must be:
(a) Processed fairly and lawfully;
(b) Collected for specified, explicit and legitimate purposes and not further processed in a way incompatible with those purposes…;
(c) Adequate, relevant and not excessive in relation to the purposes for which they are collected and/or further processed;
(d) Accurate and, where necessary, kept up to date (all reasonable steps should be taken to ensure that data which are inaccurate or incomplete in relation to the purposes for which they are collected or for which they are further processed, are erased or rectified;
(e) Kept in a form which permits identification of data subjects for no longer than is necessary for the purposes for which the data are collected or for which they are further processed.

Regulations may require organizations to report on their own compliance and impose financial penalties or loss of a license to operate if these reports are made incorrectly or outside of a directed schedule. The risk practitioner who is tasked with either building such a program or administering one already in place should work with legal counsel and other professionals across the organization to ensure an adequate understanding of what is expected. Reports on compliance should be accurate, complete and submitted in a timely manner. To ensure compliance, an organization must have the ability to monitor and measure the controls in use. Reports should be comparable from one reporting period to another, and any trends or areas of noncompliance should be identified and addressed. In cases where there is noncompliance, a justification of the reasons for noncompliance should be documented and provided on request.

In some cases, compliance may also apply to voluntary standards. For instance, the Payment Card Industry Data Security Standard (PCI DSS), created by members of the payment card industry, is not required by law but is generally required by the issuers of payment cards to be adopted by companies that want to handle payment cards. **Figure 1.18** lists the twelve requirements imposed by PCI DSS v3.0 as an example of the sort of requirements that a standard may impose on an organization.

Figure 1.18—PCI Data Security Standard—High Level Overview

The 12 PCI DSS Requirements	
Build and Maintain a Secure Network and Systems	**1.** Install and maintain a firewall configuration to protect cardholder data **2.** Do not use vendor-supplied defaults for system passwords and other security parameters
Protect Cardholder Data	**3.** Protect stored cardholder data **4.** Encrypt transmission of cardholder data across open, public networks
Maintain a Vulnerability Management Program	**5.** Protect all systems against malware and regularly update anti-virus software or programs **6.** Develop and maintain secure systems and applications
Implement Strong Access Control Measures	**7.** Restrict access to cardholder data by business need to know **8.** Identify and authenticate access to system components **9.** Restrict physical access to cardholder data
Regularly Monitor and Test Networks	**10.** Track and monitor all access to network resources and cardholder data **11.** Regularly test security systems and processes
Maintain an Information Security Policy	**12.** Maintain a policy that addresses information security for all personnel

Provided courtesy of PCI Security Standards Council, LLC and/or its licensors. ©2006-2015 PCI Security Standards Council, LLC. All Rights Reserved. *Payment Card Industry (PCI) Data Security Standard,* v3.0

The risk practitioner should always keep in mind that compliance is a risk decision, even when it is imposed by force of law. As discussed earlier, risk is addressed in the most cost-effective manner possible based on the risk appetite set by senior management. Depending on the penalties attached to noncompliance, an organization may choose not to be compliant with certain laws or regulations if the cost of compliance is greater than the fine or consequences imposed for failure to comply.

1.5.7 Establishing an Enterprise Approach to Risk Management

Risk management is an enterprise activity that benefits from a standardized, structured approach that can be applied to the entire enterprise without substantial modification or customization. It is possible to identify risk on a system-by-system or project-by-project basis, but the result of such an approach actually creates new risk of false assurance by having neither consistency nor interoperability among the risk solutions that are implemented. Without a structured approach, risk may be measured differently in different areas, creating gaps between boundaries of the various projects or systems.

This is not to say that risk should never be tailored. In a large enterprise consisting of many divisions, departments, lines of business and products or services, different cultures and business models may warrant tailoring to the needs of each organization. There is no one risk management approach that is suitable for all types of organizations or even for all components of a large, diverse organization. The risk practitioner must be sensitive to local departmental cultures, priorities, regulations, goals and restraints before recommending a risk management approach or framework. However, the recommended framework should remain substantially consistent across the organization even when it is tailored. In particular, it is important that the results of risk management in one division be comparable to the results in another.

In some large organizations and governments, the risk management function serves a consultative and advisory role that provides data and recommendations to each department but does not actually determine the development of the control or risk mitigation framework. In such an organization, risk management is a clearinghouse for information that can be accessed by all departments as a part of systems or project development. This approach is legitimate and can be effective when risk is generally integrated throughout the culture of the organization.

Executive Sponsorship (Tone at the Top)

Enterprise risk management benefits strongly from the clear support from senior management, which should require consultation with risk practitioners to be part of any new project and ensure that recommendations of the risk management program are evaluated and objectively addressed before approving or funding projects or business initiatives.

Policy

A critical part of establishing the risk management process is the development and approval of a concise, coherent risk management policy that reflects the attitude and intent of management in relation to risk. A risk management policy should include a statement relating to the reasoning or rationale behind the approach to accepting or mitigating risk, set accountability, and articulate a commitment to continuous improvement of the risk environment.

1.6 IT CONCEPTS AND AREAS OF CONCERN FOR THE RISK PRACTITIONER

The risk practitioner is not expected or required to be a technical expert in the design, implementation or support of IT systems and applications. However, strong knowledge of general IT concepts is invaluable for anyone whose roles and responsibilities require close and ongoing interaction with IT staff. The identification, assessment and monitoring of IT risk, as well as recommendation appropriate responses to the risk that is identified, require that type of close and ongoing interaction.

In each of the following sections, particular areas of concern are identified for the risk practitioner. These may encompass threats, vulnerabilities or business impacts, and this material is presented in the context of the risk management process.

1.6.1 Hardware

IT systems and business processes rely on hardware—the equipment and devices that process, store and transmit data. Hardware includes many devices, such as:
- Central processing units (CPUs)
- Motherboards
- Random access memory (RAM)
- Read-only memory (ROM)
- Networking components (switches, routers, etc.)
- Firewalls and gateways
- Keyboards
- Monitors

Areas of concern associated with hardware include:
- Obsolescence
- Poor maintenance
- Misconfiguration
- Lack of documentation
- Loss or theft
- Data loss due to insecure disposal
- Sniffing (capturing traffic)
- Physical access
- Failure
- Use without authorization

1.6.2 Software

Software includes the applications, operating systems (OSs), utilities, drivers, middleware, application program interfaces (APIs), database management systems (DBMS) and network operating systems that manage data, interface between systems, provide a user interface to hardware, and process transactions on behalf of the user.

Areas of concern associated with software include:
- Logic flaws or semantic errors
- Bugs (semantic errors)
- Lack of patching
- Lack of access control
- Disclosure of sensitive information
- Improper modification of information
- Loss of source code
- Lack of version control
- Lack of input and output validation

Operating Systems

The OS is the core software that allows the user to interface with hardware and manages all system operations. The OS manages access to system resources and controls the behavior of system components.

Areas of concern associated with the OS include:
- Unpatched vulnerabilities
- Poorly written code (buffer overflows, etc.)
- Complexity
- Misconfiguration
- Weak access controls
- Lack of interoperability
- Uncontrolled changes

Applications

Applications are the face of the information system and are the mechanism by which most users can access information, perform transactions and use system features.

Areas of concern associated with applications include:
- Poor or no data validation
- Exposure of sensitive data (i.e., lack of encryption/obfuscation)
- Improper modification of data
- Logic flaws (i.e., logic errors)
- Software bugs (i.e., coding errors)
- Lack of logs
- Lack of version control
- Loss of source code
- Weak or lack of access control

- Lack of operability with other software
- Back doors
- Poor coding practices

Software Utilities

Areas of concern associated with utilities and drivers include:
- Use of outdated drivers
- Unavailability of drivers
- Unpatched drivers
- Use of insecure components (older encryption algorithms)
- Unpatched vulnerabilities

1.6.3 Environmental Controls

Environmental controls include power and heating, ventilation and air conditioning (HVAC) systems that support IT operations. Areas of concern associated with environmental controls include:
- Power interruptions
  - Loss of power
    - Surge
    - Spikes
    - Sags
    - Brownouts
    - Faults
  - Generators
    - Insufficient capacity
    - Poor maintenance
  - Batteries
    - Poorly maintained
    - Outdated
- HVAC
  - Overheating
  - Humidity problems
    - Corrosion and condensation (high humidity)
    - Static (Low humidity)
  - Clogged filters
  - Lack of maintenance
- Water
  - Loss of water (needed for cooling systems)
- Health and safety issues
- Secure operational areas
  - Restricted access to server rooms and wiring closets
  - Secure access to power supplies, generators, elevator shafts

1.6.4 Platforms

IT platforms vary by organization. Less complex companies may have all of their technology residing on a single platform such as the mainframe, while other organizations utilize a variety of platforms to capture, store and process information. An organization's infrastructure may be located in a single centralized location or may be decentralized in multiple locations. It may be managed by its own staff or outsourced to a service provider located within the organization or at the provider's facility.

Some organizations use a multi-tiered platform, such as a three-tiered infrastructure that uses middleware, to interface between legacy systems and a user-friendly front-end application. Many systems integrate with databases that may be located on different networks or use virtual environments or thin client infrastructures that reduce functionality and vulnerabilities at the client machine level.

An example of centralized architecture, shown in **figure 1.19**, is the use of a mainframe in which each user connects to a common processer—in this case a mainframe. The users may connect through a dumb terminal or a personal computer. In a traditional mainframe, all of the processing was done on the mainframe and the user's computer was only used for data entry and display. Currently, some pre-processing is done on the user's computer, including input validation and generating the graphical display to make the data more presentable to the user.

Figure 1.19—Centralized Architecture—Mainframe

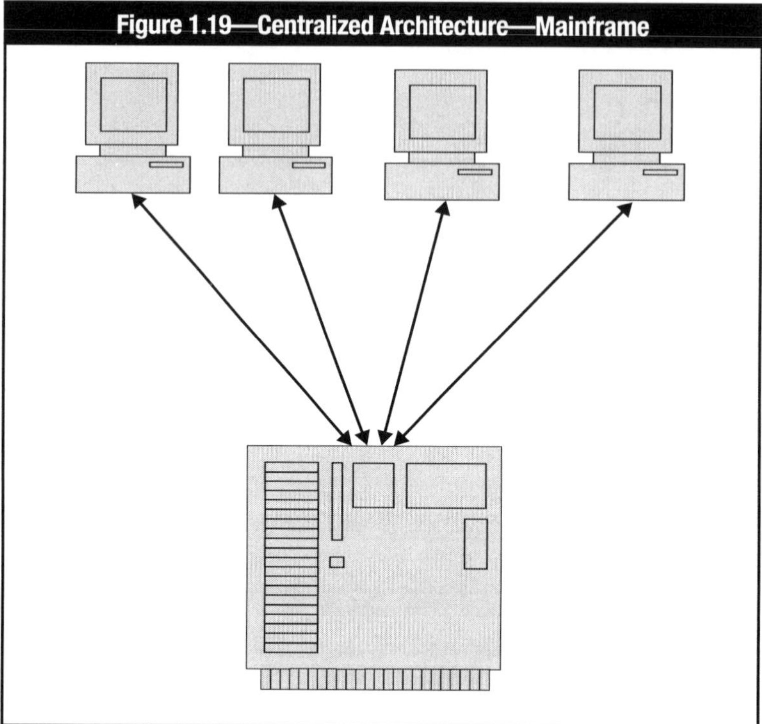

A three-tiered architecture, shown in **figure 1.20**, may use middleware to provide an interface between the user and numerous underlying systems that may be located on various servers. This allows the user to have one interface with several underlying servers. This provides the user with a much easier method of accessing data from disparate systems without having to log on to each system individually.

Figure 1.20—Three-tiered Architecture

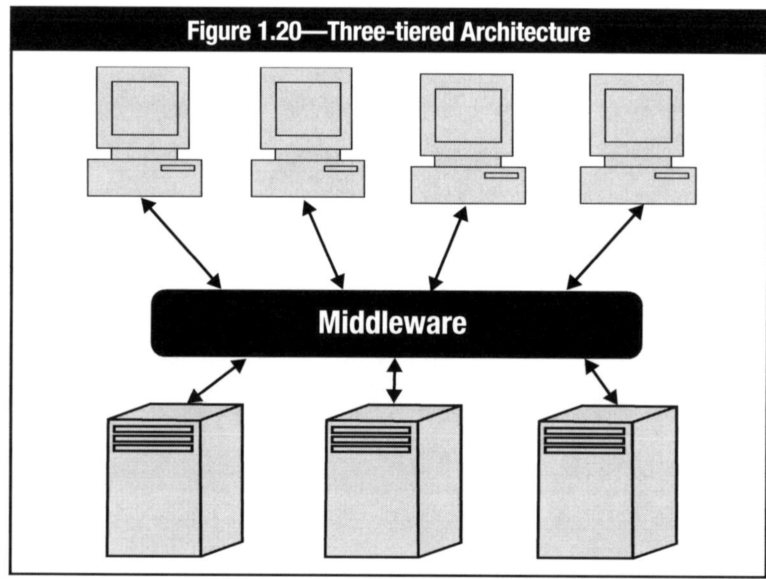

The choice of platform or platforms that are in use is not as important as the way in which the platform(s) is managed. Some platforms are more secure than others, such as a mainframe, while others are more vulnerable, such as a web application or a kiosk operating in a public area. In some cases, a system is deployed into a vulnerable environment or a legacy system is still in use that does not have sufficient native controls. In these cases, the risk practitioner should look for compensating controls that can be used to provide additional protection for the systems.

The risk practitioner must assess the risk associated with the attitude and diligence displayed by the architects and IT operations staff to harden systems, follow good practices in change control while performing administrative functions and ensure that the systems are tested on a scheduled basis to ensure compliance with standards, procedures and good security practices.

1.6.5 Network Components

A network is made up of many devices including cabling, repeaters, switches, routers, firewalls, gateways and wireless access points. Most modern networks are digital, where the communications are sent via digital signals such as bits or pulses of light. Digital signals tend to be much higher quality than earlier analog technologies based on sound waves, but they suffer from more attenuation (loss over distance) than analog traffic and may require repeaters to regenerate the signal and extend its range. Fortunately, repeaters can rejuvenate even the faintest digital signal, whereas analog attenuation tends to introduce noise that makes rejuvenation difficult.

Networks are an important part of IT risk management because they are often targets or channels used to attack systems or applications. The network itself is often essential to business operations, and many business processes rely on the availability of the network. The simple definition of a network is a system of interconnected computers and the communication equipment used to connect them. Networks may range in size from a small personal area network of a headset talking to a smartphone over Bluetooth®, to a global area network that links devices together worldwide, such as the Internet.

Networks are used for many purposes, including:
- Transferring data between individuals
- Transferring data between applications
- Controlling and monitoring of remote equipment (supervisory control and data acquisition [SCADA] networks)
- Backing up data (storage area networks [SANs])
- Enabling communication between devices (Bluetooth)

When assessing networks, the risk practitioner should consider:
- Network configuration and management, including the recognition of the criticality of network operations
- Network equipment protection
- The use of layered defense (defense in depth)
- Suitable levels of redundancy
- Availability of bandwidth
- Use of encryption for transmission of sensitive data
- Encryption key management
- Use of certificates to support PKI
- Damage to cabling and network equipment
- Tapping network connections and eavesdropping on communications
- Choice of network architecture
- Documentation of network architecture

Cabling

Networks may be connected using cable or wireless technologies. Cabling comes in several forms including unshielded twisted pair (UTP), coaxial or fiber. Each type of cable has its own benefits and vulnerabilities, and the risk practitioner should ensure that the appropriate cabling is being used for the communications required.

The least expensive option used for many local area networks (LANs) is UTP cable with a grade of category 5e (CAT5e) or category 6 (CAT6). Category (CAT7) is available, but not yet in common use. CAT7 cable is a shielded cable that protects each pair of wires and the cable itself, thereby reducing noise and cross talk for ultra-high speed

Ethernet. CAT5e and CAT6 cables are relatively easy-to-use and inexpensive cabling options that support fast Ethernet.

Areas of concern for the risk practitioner regarding cabling include:
- Physical security of cabling
- Cable exceeding the approved length of the cable runs (100 meters for CAT5e, 55 meters for CAT6)
- Protection from damage to cabling (conduit)
- Use in an area of high radio frequency interference (RFI) (may require shielding)
- Use of cable that is not of suitable standard (e.g., CAT3)
- Ensuring use of plenum-rated cable where required
- Improper terminations of cable on connectors
- Lack of cabling records

Repeaters

Repeaters are used to extend the length of a signal being transmitted over cable or wireless networks. As the signal traverses the network, it suffers from attenuation, and the repeater will be placed at a location to receive the input signal and regenerate the signal and send it on further. The advantage of a repeater is that it can filter out some noise or errors that may be affecting traffic. The distance between repeaters is based on the type of cabling or technology being used and the operational environment. For example, as the quality of fiber has increased over the years, it has meant that the distance between repeaters has grown substantially. In an environment with many concrete walls or other interference or where a higher frequency wireless channel is being used, repeaters may be required more often.

A risk associated with repeaters is ensuring that enough repeaters are in use to provide a clean, error-free signal. Another risk is a wireless repeater providing a strong signal into areas outside the perimeter of the organization's facilities that could allow unauthorized access.

Switches

Several terms used in network communications are quite broad, such as switches, firewalls and virtual private networks (VPNs). There are many types of switches with various functions, just as there are several generations of firewalls—many of which are considerably different in their function and operation.

Switches are used to connect devices together, and current switches perform the functions formerly provided by hubs and bridges (layer two). Switches forward packets to a destination, can perform routing functions (a layer three switch), address translation and balancing (layer four), and perform load balancing (a layer seven switch).

Switches can be used to connect networks but also segment and divide networks through configurations such as a virtual LAN (VLAN), where different devices physically connected onto one switch may be set up to act as completely separate networks.

Switches can be wired or wireless, small and inexpensive, or large and very expensive. Each organization must determine what equipment is suitable to meet its network requirements.

Areas of concern associated with switches include:
- Physical protection of the switch
- Ensuring the proper configuration of the switch
- Documentation
- Being a single point of failure

Routers

The purpose of a router is to connect multiple networks together and forward incoming packets in the direction of the destination Internet Protocol (IP) address that is in the packet header. A router can vary in size, from a wireless router in a home that connects computers to the Internet, to large core routers that are responsible for the routing of high-speed Internet communications. The router builds a routing table using routing protocols, such as Border Gateway Protocol (the first routers were called gateways), and learns how best to direct traffic toward its destination. In many ways, a router is like a traffic circle used to handle vehicular traffic. Traffic enters the circle and then routes off on the road that is best toward the vehicle's ultimate destination. The intelligence of the routing decision is similar to the decision made by a driver—to take the shortest route, the fastest route, the route with least construction delays, etc.; therefore, a router may direct traffic over different routes depending on traffic volumes and congestion.

Just like on the road, some traffic has labels that indicate it has a priority, such as emergency vehicles that are permitted to bypass other traffic. The label, like the siren on an emergency vehicle, indicates that this packet should be processed quickly and not be held back by other packets waiting to be processed. The delay in processing is called latency, and this would severely affect the quality of some types of traffic such as Voice-over IP (VoIP). The labeling of the traffic allows for provision of quality of service (QoS) and class of service (CoS) traffic management. QoS usually is used to provide a guaranteed bandwidth for traffic volumes, and CoS is used to grant priority to certain packets over others (e.g., voice packets over regular data packets). This also prevents the jitter—the variation in the arrival time of a packet due to different routing, latency, etc.—that would affect data traffic from impacting voice communications.

Areas of concern associated with a router include:
- Improper configuration
- Use of weak protocols (e.g., RIPv1)
- Software bugs
- Unpatched systems
- Physical security
- Unintentional support for IPv6 (e.g., Teredo tunneling capabilities turned on by default)

Firewalls
A firewall is defined as a system or combination of systems that enforces a boundary between two or more networks, typically forming a barrier between a secure and an open environment, such as the Internet. The risk practitioner must remember that firewalls should be used to control both incoming and outgoing traffic. The term "firewall" refers to many types of technology that operate at various layers of the open systems interconnection (OSI) model and have evolved significantly over the past years. Types of firewalls are described in **figure 1.21**

Figure 1.21—Firewall Types	
First Generation	A simple packet-filtering router that examines individual packets and enforces rules based on addresses, protocols and ports.
Second Generation	Keeps track of all connections in a state table. This allows it to enforce rules based on packets in the context of the communications session.
Third Generation	Operates at layer seven (the application layer) and is able to examine the actual protocol being used for communications, such as Hypertext Transfer Protocol (HTTP). These firewalls are much more sensitive to suspicious activity related to the content of the message itself, not just the address information.
Next Generation	Sometimes called deep packet inspection—is an enhancement to third generation firewalls and brings in the functionality of an intrusion prevention system (IPS) and will often inspect Secure Sockets Layer (SSL) or Secure Shell (SSH) connections.

It is important to ensure that the firewall configurations are backed up on a regular basis, reviewed to ensure that all rules are in the correct order and are documented, and tested on a scheduled basis. Any changes to the firewall configuration or rules should be subject to the change management process of the organization.

A firewall is only as good as the person that manages it. An important role of the risk practitioner is to verify that the staff members managing the firewalls and other network devices are knowledgeable, trained and supervised. Firewall logs must be reviewed on a regular basis to detect any suspicious activity.

Proxy
A proxy is a device that acts as an intermediary between two communicating parties. The proxy acts to the party on each side of the communication as if it were the other end host. This allows it to filter and examine suspicious activity, protect internal resources and take action if unacceptable activity is occurring. A gateway is a type of proxy that controls traffic through a gate or security perimeter.

Domain Name System

For most people, the domain name system (DNS) is the mechanism that makes the Internet work. Without the DNS, they cannot navigate to their destination web site without knowing the IP address of the web site they wanted to visit. The DNS provides a simple cross-reference that is used to associate a normal name with the IP address used by network devices. The IP address is a logical address given to a web site based on the addresses allocated to the Internet service provider (ISP). An individual can purchase a name for his/her web site that is more in line with his/her name, the name of the organization or the marketing program. For example, if the IP address for the ISACA home page were 192.168.1.0, a person wanting to visit the ISACA web site could type that number and reach the site, but he/she would probably find it easier to remember *www.isaca.org*.

DNS is what makes this possible. DNS is a tree structure that resolves the addresses so that a network device that receives a request from a person entering *www.isaca.org* will be able to look up the IP address and then route the request properly through the network. When a network device does not know the IP address associated with a web name, it sends a DNS request up through the DNS tree to a higher level DNS resolver. The reply is then sent to the requesting device using User Datagram Protocol (UDP) over port 53, after which the requesting device stores the IP address and name to reference for future use.

There have been many attacks on the Internet using DNS over the years. People have sent false DNS replies to misroute traffic. DNS replies have been used in amplification attacks to flood a victim's system. DNS can also be used to learn information about a company that may be useful in planning attacks.

Wireless Access Points

Wireless devices offer a level of flexibility and ease of use not possible with traditional cable-based networks, allowing users to move around and log in from multiple locations without requiring network cables or ports. However, the lack of a defined physical boundary increases the threat of unauthorized people being able to access the network. Good practice is to segment the wireless infrastructure into a separate network and implement strong password requirements for network access. Placement of wireless access points in locations not subject to interference from other devices or near a window that would broadcast a strong signal outside of the organization's facilities is also important.

The installation of rogue or unauthorized wireless access points on the network may permit a person to access the network directly without having to connect through a secure device or to bypass firewalls or other layers of defense. Detecting this type of vulnerability is generally possible only with a combination of diligence and technical expertise, particularly if the rogue access point has been installed by someone who has legitimate physical access to the organization's workspace. The risk practitioner should ensure that wireless discovery scans are part of the overall vulnerability assessment process, even (especially) if wireless technology is not known to be used within the organization.

Other Network Devices

Besides the network devices listed above, there are many other types of networking equipment and protocols used for network communications. The proper configuration of these devices is a critical part of secure communications and network reliability and stability. The choice of communication protocols used is important because some protocols are more secure than others. A common problem related to network security is unpatched systems or protocols, such as the Network Time Protocol (NTP)-based attacks of late 2013 that resulted from many organizations having unpatched NTP implementations.

1.6.6 Network Architecture

Many network architectures in use today range from older bus networks to high-speed fiber backbones. The choice of network architecture is based on the number of devices (e.g., an old peer-to-peer network could only handle 8 to 12 devices), distance (e.g., LAN versus wide area networks [WANs]) and communications requirements of the network (e.g., speed, confidentiality, reliability).

Types of Network Topologies

Figure 1.22 illustrates commonly used physical network topologies.

Figure 1.22—Physical Topologies in Common Use

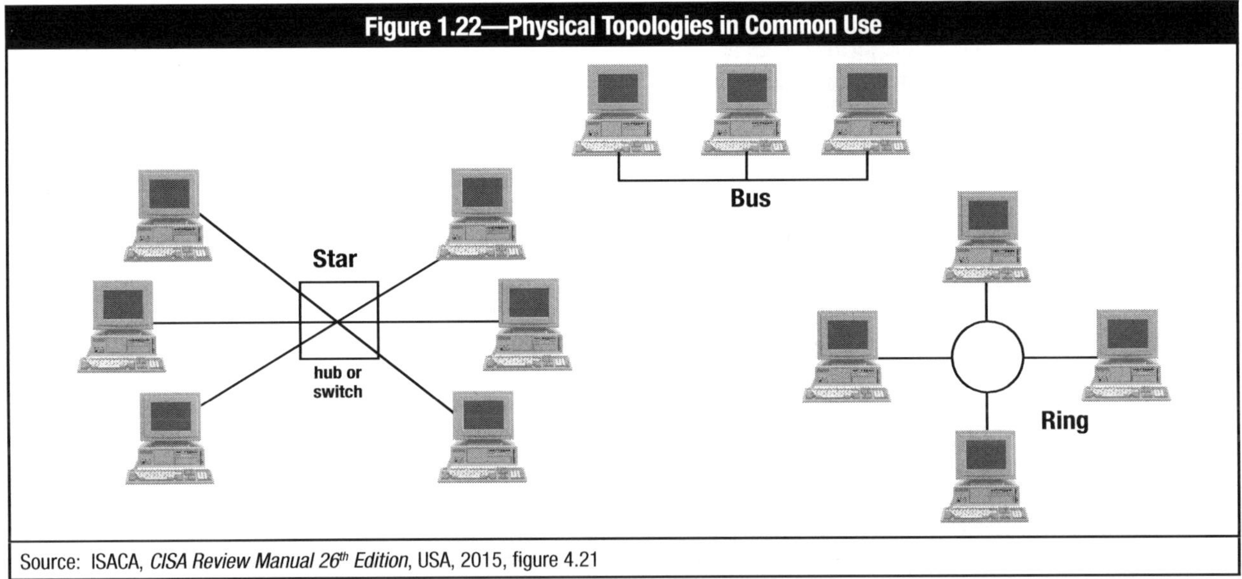

Source: ISACA, *CISA Review Manual 26th Edition*, USA, 2015, figure 4.21

A bus network topology connects every device by means of one communications path. The design is simple and cost-effective, which is why it has been used for cable television signals for decades. The key vulnerability lies in upstream dependency; a cut cable results in failure for every user connected downstream of the cut. It is also relatively easy to intercept traffic on a bus network (a point of particular concern for users of cable-based Internet access).

In a star network topology, every device is connected to a central switch, which is more efficient than a bus on account of shorter paths from any end point to the switch. The switch design also makes it more difficult for one user to intercept traffic intended for another. Star topologies are even more vulnerable to interruption than bus topologies, because loss of the central switch affects all users. However, the switch can generally be more easily protected than a bus cable.

A tree network topology is a series of star networks arranged with branches to other star networks in a tree-type structure. A cut link between the branches of the tree causes isolation of that branch. Tree networks are popular because they are very scalable.

A ring network topology is used in backbones and areas where reliable high-speed communications and fault tolerance is desired. A ring connects every device and allows traffic to pass in one or both directions; if the ring is bidirectional, a single cut does not affect the network at all. Ring topologies are excellent for busy networks where switching would introduce too much delay due to processing overhead, but they are considerably more expensive that bus, star or tree topologies.

A mesh network topology is used where high availability is required. Many devices are connected to many other devices in a mesh so that traffic can route around a failure in any part of the network. The Internet itself is a partial mesh and was built to survive massive failures of any one part of the network and still allow communications over the rest of the Internet. Mesh networks are costly, and they do not scale with physical lines due to the need to connect each device to many other devices. Wireless links make locally oriented mesh architectures considerably easier to architect and maintain.

Any network can fail, so the risk practitioner must consider the impact of any failure and the ways that a failure can be mitigated.

Local Area Network

A local area network (LAN) is defined as a communication network that serves several users within a specified geographic area, such as a building or a department. Shared data are stored in a file server that acts as a remote disk drive for all users in the network. LANs often use star or tree architectures. Multiple LANs connected together over long-haul links are referred to as wide area networks (WANs).

LANs intended purely for personal use, such as those built within homes, are called personal area networks (PANs) by some sources. Additionally, the term storage area network (SAN) refers to a variation of a LAN built for the express purpose of connecting storage devices to servers and other computing devices.

Wide Area Network

A WAN is defined as a computer network connecting different remote locations that may range from short distances, such as a floor or building, to extremely long transmissions that encompass a large region or several countries. Some sources refer to WANs that extend only over limited areas such as cities as metropolitan area networks (MANs).

There are several different topologies that can be used when a WAN must connect networks over a large geographic area, such as leased lines, packet-switching networks, satellite and wireless.

Leased Lines

A leased line is leased or rented from a telecommunications carrier provided for the sole use of the organization that leases it. By definition, a leased line provides a private network, and leased lines are available in a variety of speeds and configurations, ranging from 56 kilobits per second up to many gigabits per second for optical fiber. Leased lines are typically expensive; digital subscriber line (DSL) technology that makes use of existing copper pair telephone lines to provide service of up to several megabits per second (depending on the location) is an exception.

Packet-switching Networks

The development of packet-switching networks, in which traffic can be divided into parts that find their own paths to destinations for later reassembly, made it possible to share physical communications among multiple organizations and reduced the cost of networking considerably. Some of the technologies used in packet-switched networks are X.25, frame relay and asynchronous transfer mode (ATM). Benefits include class and quality of service (CoS and QoS) features that guarantee a certain amount of bandwidth availability to the organization and give priority to certain types of traffic. Through the implementation of multiprotocol label switching (MPLS), it is also possible to engineer the route that traffic takes through the network.

Packet-switching networks are subject to latency and jitter, and some packets can be received out of order which can impact streaming communications and affect traffic quality. Permanent virtual circuits (PVCs) that mandate the route that traffic takes through the network can reduce or prevent jitter and provide higher standard levels of communications quality. The single point of failure on a packet-switching network is the link between the organization and point at which the organization connects to the telecommunications provider.

Microwave

Microwave is a line-of-sight technology where the sending and receiving stations need to have a linear path without interruption. A building, tree, strong wind or other obstacle can affect the reliability of microwave communications. Microwave is also licensed in many countries, requiring an organization to purchase the right to broadcast, and it may be intercepted or monitored.

Optical

Optical technology is based on laser pulses and can transmit fairly high-speed communications. Like microwave, optical depends on an uninterrupted line of sight, but it is not licensed in most countries.

Satellite

Satellite communications enable transmission and reception to and from remote areas where it is not possible to provide other forms of communications. Satellite is a line-of-sight technology and can be affected by wind, heavy rain or snow, dust, or solar flares. Although satellite signals can carry large amounts of information at a time, the disadvantage is a bigger delay due to propagation delay, the "jump" from the earth to the satellite and back (approximately 300 milliseconds).

Demilitarized Zone

The area of the network that is accessible to outsiders through the Internet is isolated into a DMZ. This prevents an attacker from having direct access into internal systems. All devices in the DMZ are hardened with all unnecessary functionality disabled. Such devices are often referred to as a bastion host. Intrusion detection systems (IDSs) and intrusion prevention systems (IPSs) monitor, record and may block suspicious activity. The application firewall in the DMZ is behind a packet filtering router to clear out most of the bad traffic before it gets to the application firewall.

As can be seen in **figure 1.23**, a user from the Internet can only access the organization through the firewall. The firewall ensures that traffic from the outside is routed into the DMZ where the web server is located. Nothing valuable is kept in a DMZ because it is subject to attack and compromise from the outside. Outside users cannot access the internal network directly, thus providing the internal network with some measure of protection.

Figure 1.23—The Demilitarized Zone

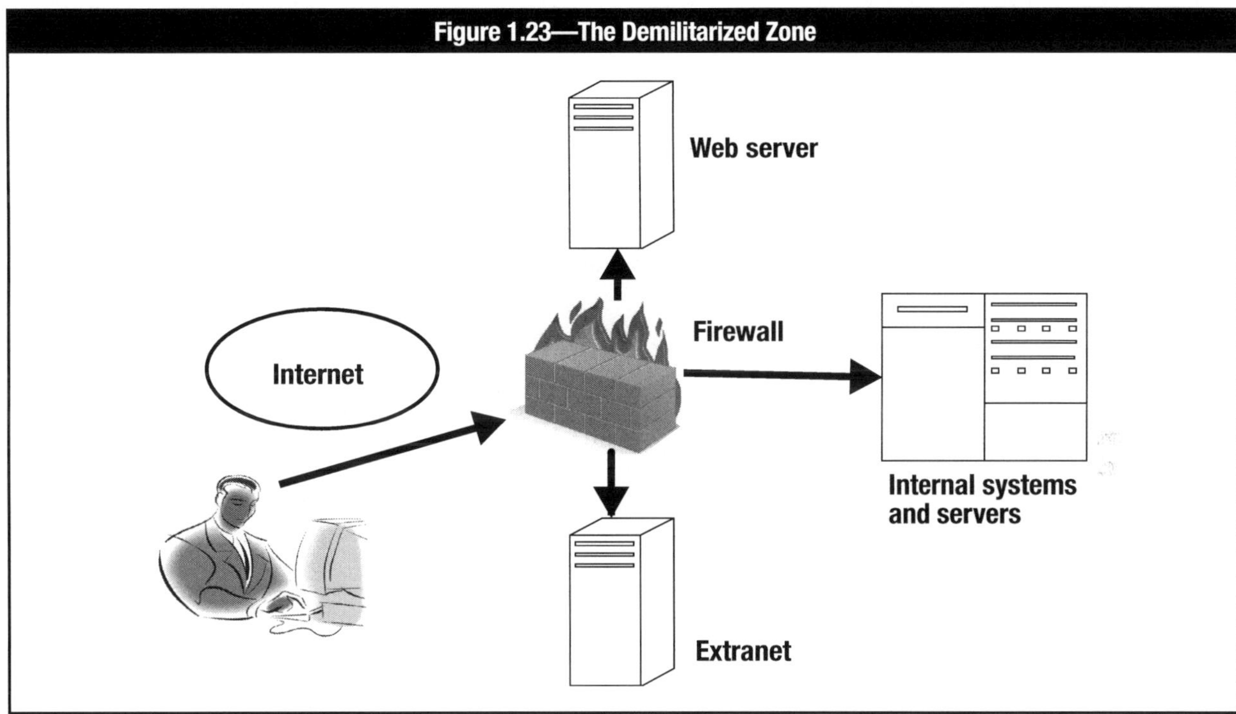

Encryption

Encryption is a vital part of secure communications and is used widely in many products and systems throughout an organization ranging from email to VPNs, access control, backup tapes and hard disk storage. The secure use of encryption requires the correct choice of algorithm, the protection of the encryption keys, the randomness of key generation and limits on the length of time a key is used before being changed. In many cases, the breach of an implementation of encryption is due to user misuse or an error in the implementation, not to a problem with the algorithm. The risk practitioner is concerned with how keys are generated and stored, the training of the users, and ensuring that the algorithms adhere to current standards. Encryption is discussed in greater detail in chapter 3 Risk Response and Mitigation.

Virtual Private Network

Under ordinary conditions, it is fairly easy to capture traffic passing over the Internet or by way of directional connections such as wireless, satellite or microwave. A VPN is a secure private network that uses the public telecommunications infrastructure to transmit data. VPNs work by digitally constructing tunnels so that content traffic within the tunnel is logically separated from other traffic. Once established, the VPN functions in a manner similar to a leased line but at a considerably lower cost.

Although it is common for VPNs to be paired with encryption, and certain technologies automatically incorporate encryption, the risk practitioner should be aware that construction of a VPN tunnel is distinct from whether or not that tunnel is encrypted. Where encryption is used, the risk practitioner should take note of whether it is permitted to traverse network boundary devices such as firewalls. It is good practice for VPNs to terminate outside of firewalls so that the traffic may be inspected.

Where many users need to connect securely, an organization may use a VPN concentrator, a dedicated system used to establish VPN tunnels and handle large numbers of simultaneous connections. This system provides authentication, authorization and accounting services.

1.7 METHODS OF RISK IDENTIFICATION

The risk practitioner has several possible sources for identification of risk, including:

- Historical- or evidence-based methods, such as review of historical events, such as:
  - Audit or incident reports
  - Public media (e.g., newspapers, television, etc.)
  - Annual reports and press releases
- Systematic approaches (expert opinion) in which a risk team examines and questions a business process in a systematic manner to determine the potential points of failure, such as:
  - Vulnerability assessments
  - Review of business continuity and disaster recovery plans
  - Interviews and workshops with managers, employees, customers, suppliers and auditors
- Inductive methods (theoretical analysis), where a team examines a process to determine the possible point of attack or compromise, such as penetration testing

The process of risk identification serves as the input to another aspect of risk management, as seen in **figure 1.24**.

Figure 1.24— Risk Identification Process

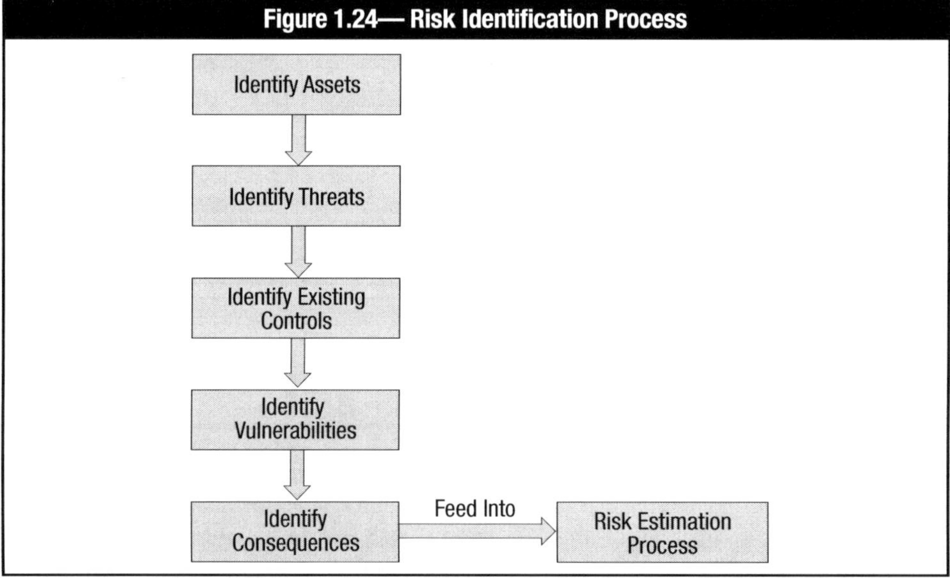

1.7.1 Conducting Interviews

Gathering information from staff is a valuable method of gaining insight into the business from those closest to the processes and most likely to understand their fundamental workings. However, interviews pose certain challenges of which the risk practitioner must be aware. One is that many people want to be seen as essential participants in the mission of the organization, which can lead to exaggeration of their own importance and that of their teams or departments; another is that people may not fully understand the overall business process or the dependencies between their department and other departments. In certain cases, such as where there is demonstrable negligence or wrongdoing, someone may intentionally provide incorrect information.

The risk practitioner can improve the odds of an interview producing useful information by adhering to the following good practices:

- Designate a specific time period and do not exceed that time without mutual agreement.
  - When a manager is told that a staff member will be needed for 45 minutes, he/she should not discover that the interview lasted 90 minutes.

- Know as much as possible about the business process in advance of the interview, which reduces time spent on general explanations of core business functions.
  - Obtain and review documentation such as process maps, standard operating procedures, the results of impact assessments and network topologies before the interview.
- Prepare questions and provide them to the interviewee in advance so that he or she can bring any supporting documentation, reports or data that may be necessary.
- Conduct interviews with senior leaders in order to ensure a thorough understanding of the enterprise, including every aspect of each business operation.
  - Senior leaders may include board members, administrators, critical third-party service providers, customers, suppliers and managers.
- Encourage interviewees to be open about challenges they face and risk that concerns them, as well as any potential missed opportunities or problems associated with their current processes, systems and services/products.
- Avoid setting incorrect expectations regarding confidentiality of interview answers.
  - People may worry about the repercussions of discussing flaws or missed opportunities.
  - Promise confidentiality only if it will actually be maintained.

One important detail to obtain during an interview is the level of impact that previous incidents have had on the organization, including how the incident was handled, results of post-incident review and root cause analysis and current status of any noted remediation activities from prior incidents, including whether the recommendations from the previous incident were addressed and resolved. The risk practitioner should be aware the people made available for interviews may not be the correct people to answer risk-related questions. When scheduling or other conflicts impede the risk management process, the risk practitioner may need to turn to the executive sponsor of the program to facilitate access to the right people in a timely manner.

1.7.2 Risk Identification and Classification Standards and Frameworks

Several good sources for risk identification and classification standards and frameworks are available to the risk practitioner. The following list is not comprehensive, and many more standards are available. However, this list may allow the risk practitioner to consider a framework or standard that would be suitable for use in his/her organization. Many countries and industries have specific standards that must be used by organizations operating in their jurisdiction. The use of a recognized standard may provide credibility and completeness for the risk assessment and management program of the organization and help ensure that the risk management program is comprehensive and thorough.

Note: The CRISC candidate will not be tested on any specific standard. The use of any standards in this review manual is for example and explanatory purposes only.

ISO 31000:2009 Risk Management—Principles and Guidelines

ISO 31000:2009 states:

> *This international standard recommends that organizations develop, implement and continuously improve a framework whose purpose is to integrate the process for managing risk into the organization's overall governance, strategy and planning, management, reporting polices, values and culture.*
>
> *Although the practice of risk management has been developed over time and within many sectors in order to meet diverse needs, the adoption of consistent processes within a comprehensive framework can help to ensure that risk is managed effectively, efficiently and coherently across an organization. The generic approach described in this standard provides the principles and guidelines for managing any form of risk in a systematic, transparent and credible manner and within any scope and context.*[10]

COBIT® 5 for Risk

COBIT 5 for Risk is described as follows:

> *COBIT 5 provides a comprehensive framework that assists enterprises in achieving their objectives for the governance and management of enterprise information technology (IT). Simply stated, COBIT 5 helps enterprises to create optimal value from IT by maintaining a balance between realising benefits and optimising risk levels and resource use. COBIT 5 enables IT to be governed and managed in a holistic manner for the entire enterprise, taking into account the full end-to-end business and IT functional areas of responsibility and considering the IT-related interests of internal and external stakeholders.*
>
> *COBIT 5 for Risk … builds on the COBIT 5 framework by focusing on risk and providing more detailed and practical guidance for risk professionals and other interested parties at all levels of the enterprise.*[11]

IEC 31010:2009 Risk Management—Risk Assessment Techniques

IEC 31010:2009 states:

> *Organizations of all types and sizes face a range of risks that may affect the achievement of their objectives.*
>
> *These objectives may relate to a range of the organization's activities, from strategic initiatives to its operations, processes and projects, and be reflected in terms of societal environmental, technological, safety and security outcomes, commercial, financial and economic measures, as well as social, cultural, political and reputation impacts.*
>
> *All activities of an organization involve risks that should be managed. The risk management process aids decision making by taking account of uncertainty and the possibility of future events or circumstances (intended or unintended) and their effects on agreed objectives.*[12]

ISO/IEC 27001:2013 Information Technology—Security Techniques—Information Security Management Systems—Requirements

ISO 27001:2013 states:

> *The organization shall define and apply an information security risk assessment process that: c) identifies the information security risks:*
>
> *1) apply the information security risk assessment process to identify risks associated with the loss of confidentiality, integrity and availability for information within the scope of the information security management system; and*
>
> *2) identify risk owners.*[13]

ISO/IEC 27005:2011 Information Technology—Security Techniques—Information Security Risk Management

ISO/IEC 27005 states:

> *This international standard provides guidelines for information security risk management in an organization, supporting in particular the requirements of an information security management system (ISMS) according to ISO/IEC 27001. However, this standard does not provide any specific methodology for information security risk management. It is up to the organization to define their approach to risk management, depending for example on the scope of the ISMS, context of risk management, or industry sector. A number of existing methodologies can be used under the framework described in this International standard to implement the requirements of an ISMS.*[14]

NIST Special Publications

NIST has a wide range of special publications available at *csrc.nist.gov*. Some of the publications related to IT risk are discussed in the following sections.

NIST Special Publication 800-30 Revision 1: Guide for Conducting Risk Assessments

NIST Special Publication 800-30 Revision 1 describes risk assessment in the following manner:

> *Risk assessments are a key part of effective risk management and facilitate decision making at all three tiers in the risk management hierarchy including the organization level, mission/business process level, and information system level.*

Because risk management is ongoing, risk assessments are conducted throughout the system development life cycle, from pre-system acquisition (i.e., material solution analysis and technology development), through system acquisition (i.e., engineering/manufacturing development and production/deployment), and on into sustainment (i.e., operations/support).[15]

NIST Special Publication 800-39: Managing Information Security Risk

NIST Special Publication 800-39 states:

The purpose of Special Publication 800-39 is to provide guidance for an integrated, organization-wide program for managing information security risk to organizational operations (i.e., mission, functions, image, and reputation), organizational assets, individuals, other organizations, and the Nation resulting from the operation and use of federal information systems. Special Publication 800-39 provides a structured, yet flexible approach for managing risk that is intentionally broad-based, with the specific details of assessing, responding to, and monitoring risk on an ongoing basis provided by other supporting NIST security standards and guidelines.[16]

1.7.3 Example of a Risk Management Program Based on ISO/IEC 27005

Especially in cases where good risk management practices are not currently in place, the risk practitioner may wish to formally adopt or informally use one or more standard or framework to guide his/her risk management program. The program encompasses the goals of the organization as a whole, not specifically the roles and duties of the risk practitioner.

Risk management includes several components, and although the specific names of each component may differ from one standard or framework to the next, the typical progression is to first identify areas of risk and analyze them, then use the results of these analyses to choose appropriate ways of responding to them, after which the organization monitors the effectiveness of its implemented choices and reports progress to senior management. **Figure 1.25** illustrates one such program based on ISO/IEC 27005:2011.

Note: The following is provided as an example of a risk management process. The risk practitioner may find that he/she will need to adopt one or more well-established standards or frameworks to ensure that the risk management program in his/her organization is complete and authoritative.

Figure 1.25—Information Security Risk Management Process

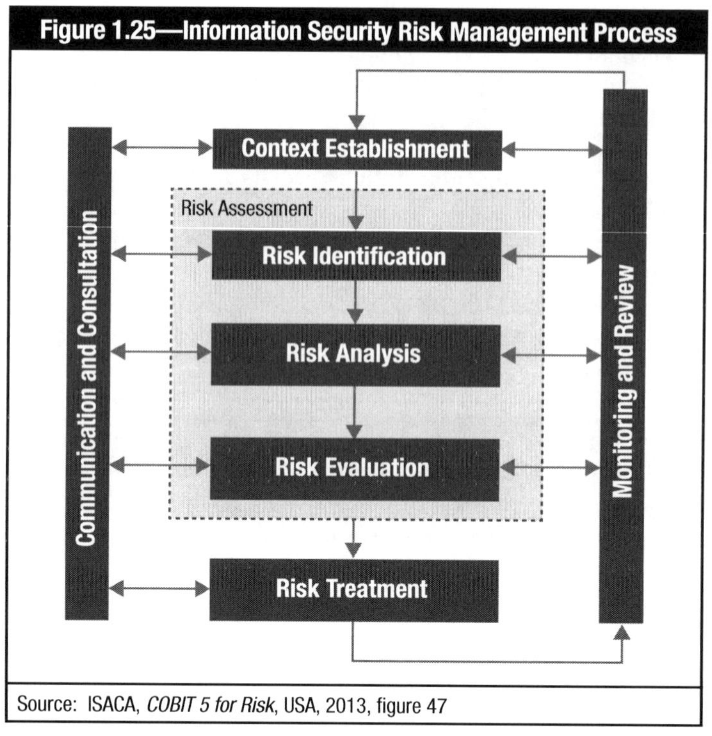

Source: ISACA, *COBIT 5 for Risk*, USA, 2013, figure 47

Figure 1.26 provides a summary of the important concepts of these process steps.

Figure 1.26—ISO/IEC 27005 Process Steps

ISO/IEC 27005 Process Step	Important Concepts of the Component
Context Establishment	This process step includes: • Setting the basic criteria necessary for establishment of information security risk management (ISRM) • Defining the scope and boundaries • Establishing an appropriate organization operating the ISRM
Risk Assessment	Risk assessment determines the value of the information assets, identifies the applicable threats and vulnerabilities that exist (or could exist), identifies the existing controls and their effect on the risk identified, determines the potential consequences, and finally, prioritizes the derived risk and ranks it against the risk evaluation criteria set in the context establishment. This process step consists of risk identification, risk analysis and risk evaluation.
Risk Identification	Risk identification includes the identification of: • Assets • Threats • Vulnerabilities • Existing controls • Consequences The output of this process is a list of incident scenarios with their consequences related to assets and business processes.
Risk Analysis	The risk estimation step includes: • Assessment of consequences • Assessment of incident likelihoods • Determination of level of risk
Risk Evaluation	In this step, levels of risk are compared according to risk evaluation criteria and risk acceptance criteria. The output is a prioritized list of risk elements and the incident scenarios that lead to the identified risk elements.

Figure 1.26—ISO/IEC 27005 Process Steps *(cont.)*	
ISO/IEC 27005 Process Step	**Important Concepts of the Component**
Risk Treatment	Risk treatment options include: • Risk modification • Risk retention • Risk avoidance • Risk sharing
Risk Acceptance	The input is a risk treatment plan and the residual risk assessment subject to the risk acceptance criteria. This stage comprises the formal acceptance and recording of the suggested risk treatment plans and residual risk assessment by management, with justification for those that do not meet the enterprise's criteria.
Risk Communication and Consultation	This is a transversal process where information about risk should be exchanged and shared between the decision maker and other stakeholders through all the steps of the risk management process.
Risk Monitoring and Review	Risk and its influencing factors should be monitored and reviewed to identify any changes in the context of the organization at an early stage and to maintain an overview of the complete risk picture.

1.8 IT RISK SCENARIOS

A risk scenario is a description of a possible event whose occurrence will have an uncertain impact on the achievement of the enterprise's objectives, which may be positive or negative.[17] The development of risk scenarios provides a way of conceptualizing risk that can aid in the process of risk identification. Scenarios are also used to document risk in relation to business objectives or operations impacted by events, making them useful as the basis for quantitative risk assessment. Each identified risk should be included in one or more scenarios, and each scenario should be based on an identified risk.

1.8.1 Risk Scenario Development Tools and Techniques

The development of risk scenarios is based on describing a potential risk event and documenting the factors and areas that may be affected by the risk event. Each scenario should be related to a business objective or impact. Risk events may include system failure, loss of key personnel, theft, network outages, power failures, natural disasters or any other situation that could affect business operations and mission. The key to developing effective scenarios is to focus on real and relevant potential risk events. Examples of this would be to develop a risk scenario based on a radical change in the market for an organization's products, a change in government or leadership or a supply chain failure.

The development of the risk scenarios purely from imagination is an art that often requires creativity, thought, consultation and questioning. Incidents that have occurred previously may be used as the basis of risk scenarios with far less effort put into their development. Risk scenarios based on past events should be fully explored to ensure that similar situations do not recur in ways that might have been avoided.

Risk scenarios can be developed from a top-down perspective driven by business goals or from a bottom-up perspective originating from hypothetical scenarios, as shown in **figure 1.27**.

Figure 1.27—Risk Scenario Overview

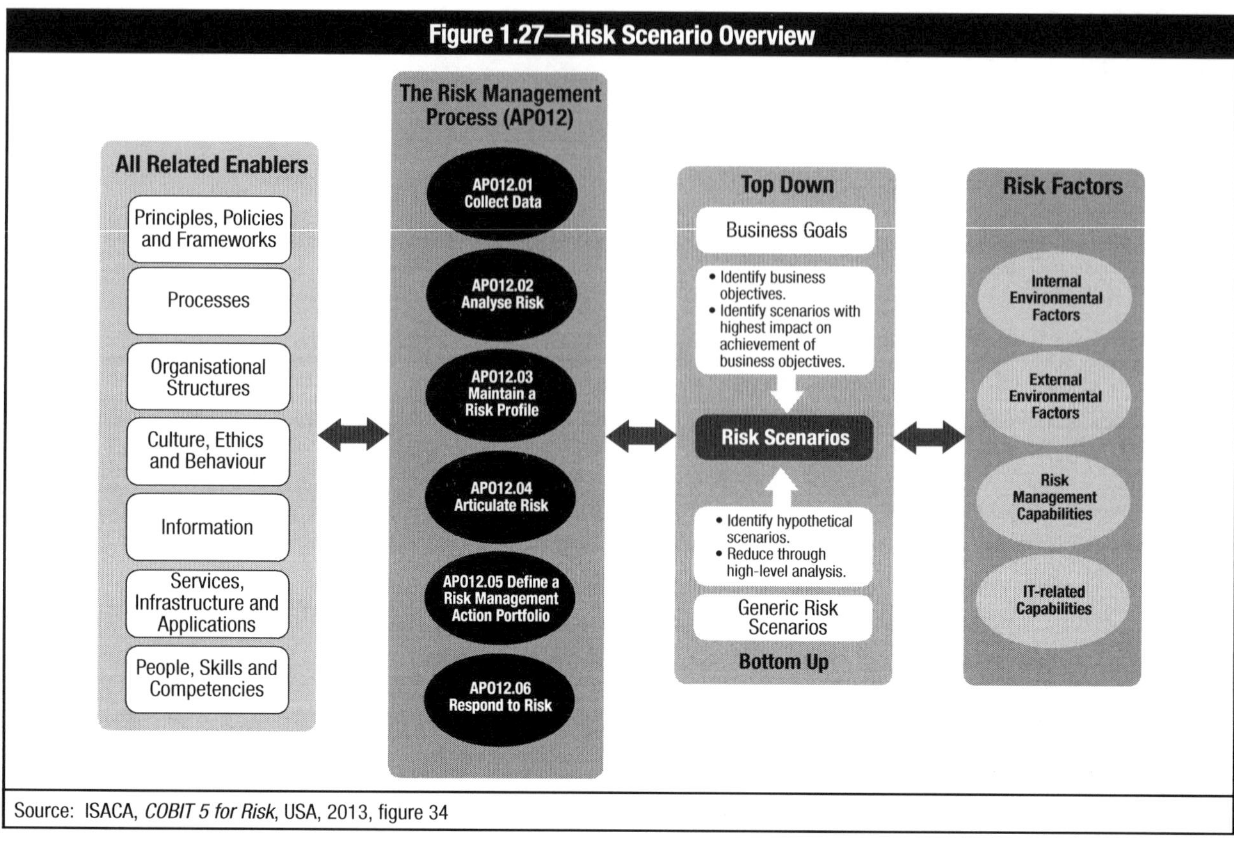

Source: ISACA, *COBIT 5 for Risk*, USA, 2013, figure 34

Figure 1.28 shows an example of the many inputs that are required to develop risk scenarios.

Figure 1.28—Risk Scenario Structure

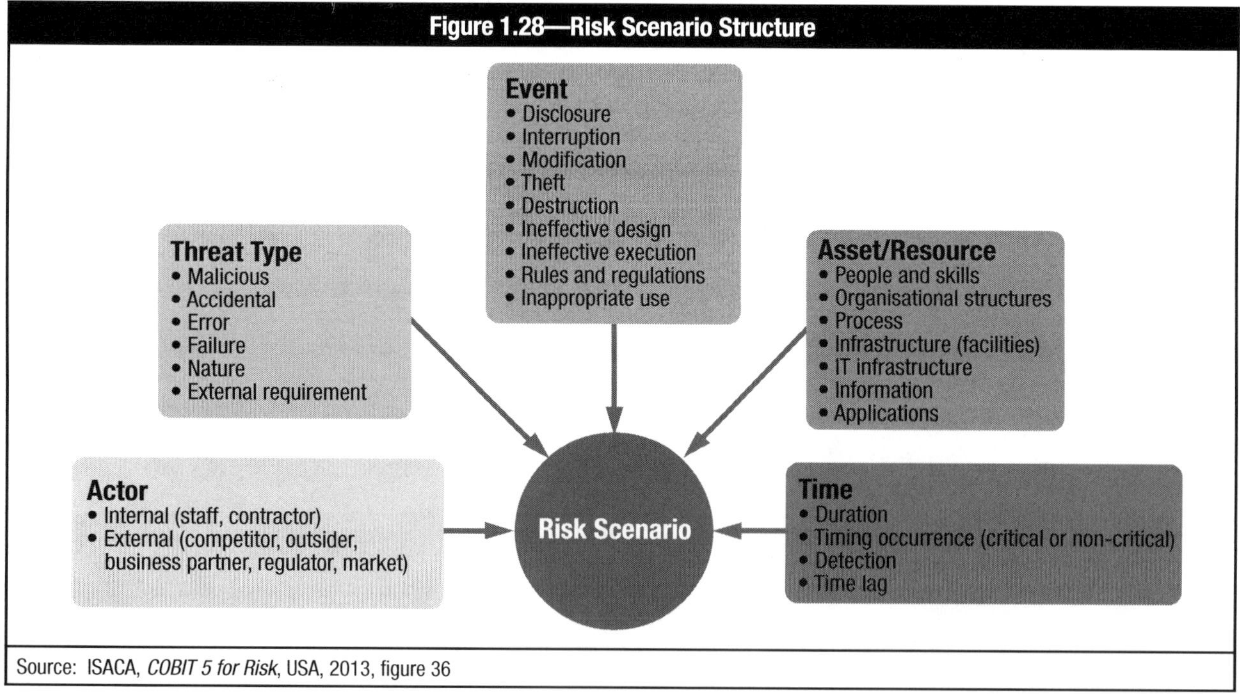

Source: ISACA, *COBIT 5 for Risk*, USA, 2013, figure 36

Top-down Approach

A top-down approach to scenario development is based on understanding business goals and how a risk event could affect the achievement of those goals. Under this model, the risk practitioner looks for the outcome of events that may hamper business goals identified by senior management. Various scenarios are developed that allow the organization to examine the relationship between the risk event and the business goals, so that the impact of the risk event can be measured. By directly relating a risk scenario to the business, senior managers can be educated and involved in how to understand and measure risk.

The top-down approach is suited to general risk management of the company, because it looks at both IT- and non-IT-related events. A benefit of this approach is that because it is more general, it is easier to achieve management buy-in even if management usually is not interested in IT. The top-down approach also deals with the goals that senior managers have already identified as important to them.

Bottom-up Approach

The bottom-up approach to developing risk scenarios is based on describing risk events that are specific to individual enterprise situations, typically hypothetical situations envisioned by the people performing the job functions in specific processes. The risk practitioner and assessment team start with one or more generic risk scenarios then refine them to meet their individual organizational needs including building complex scenarios to account for coinciding events.

Bottom-up scenario development can be a good way to identify scenarios that are highly dependent on the specific technical workings of a process or system, which may not be apparent to anyone who is not intimately involved with that work but could have substantial consequences for the organization. One downside of bottom-up scenario development is that it may be more difficult to maintain management interest in highly specialized technical scenarios.

1.8.2 Benefits of Using Risk Scenarios

Risk scenarios facilitate communication in risk management by constructing a narrative that can inspire people to take action. The use of risk scenarios can enhance the risk management effort by helping the risk team to understand and explain risk to the business process owners and other stakeholders. Additionally, a well-developed scenario provides a realistic and practical view of risk that is more aligned with business objectives, historical events and emerging threats envisioned by the organization than would be found by consulting a broadly applicable standard or catalog of controls. These benefits make risk scenarios valuable as means of gathering and framing information used in subsequent steps in the risk management process.

1.8.3 Developing IT Risk Scenarios

A risk scenario is a description of an IT-related risk event that can lead to a business impact. The risk scenario contains the following components:

- Agent: The internal or external party or entity that generates the threat
- Threat type: The nature of the threat event (malicious or accidental; a natural event; an equipment or process failure)
- Event: The security incident, such as the disclosure of information, the interruption of a system or project, including:
  - Theft
  - Improper modification of data or a process
  - Inappropriate use of resources
  - Changes to regulations
  - Lack of change management
- Asset: The entity affected by the risk event, including:
  - People
  - Organizational structure
  - IT processes
  - Physical infrastructure
  - IT infrastructure
  - Information, applications
- Time: If relevant to the scenario, including:
  - Duration (extended outage)
  - Timing (at a critical moment)
  - Detection (immediate detection or not)
  - Time lag between the event and consequence (immediate impact of an network failure versus long term problems from poor infrastructure)

Some of the key issues related to the development of risk scenarios are addressed in **figure 1.29**.

Figure 1.29—Risk Scenario Technique Main Focus Areas

Focus/Issue	Summary Guidance
Maintain currency of risk scenarios and risk factors.	Risk factors and the enterprise change over time; hence, scenarios will change over time, over the course of a project or over the evolution of technology. For example, it is essential that the risk function develop a review schedule and the CIO works with the business lines to review and update scenarios for relevance and importance. Frequency of this exercise depends on the overall risk profile of the enterprise and should be done at least on an annual basis, or when important changes occur.
Use generic risk scenarios as a starting point and build more detail where and when required.	One technique of keeping the number of scenarios manageable is to propagate a standard set of generic scenarios through the enterprise and develop more detailed and relevant scenarios when required and warranted by the risk profile only at lower (entity) levels. The assumptions made when grouping or generalising should be well understood by all and adequately documented because they may hide certain scenarios or be confusing when looking at risk response. For example, if 'insider threat' is not well defined within a scenario, it may not be clear whether this threat includes privileged and non-privileged insiders. The differences between these aspects of a scenario can be critical when one is trying to understand the frequency and impact of events, as well as mitigation opportunities.
Number of scenarios should be representative and reflect business reality and complexity.	Risk management helps to deal with the enormous complexity of today's IT environments by prioritising potential action according to its value in reducing risk. Risk management is about reducing complexity, not generating it; hence, another plea for working with a manageable number of risk scenarios. However, the retained number of scenarios still needs to accurately reflect business reality and complexity.
Risk taxonomy should reflect business reality and complexity.	There should be a sufficient number of risk scenario scales reflecting the complexity of the enterprise and the extent of exposures to which the enterprise is subject. Potential scales might be a 'low, medium, high' ranking or a numeric scale that scores risk importance from 0 to 5. Scales should be aligned throughout the enterprise to ensure consistent scoring.
Use generic risk scenario structure to simplify risk reporting	Similarly, for risk reporting purposes, entities should not report on all specific and detailed scenarios, but could do so by using the generic risk structure. For example, an entity may have taken generic scenario 15 (project quality), translated it into five scenarios for its major projects, subsequently conducted a risk analysis for each of the scenarios, then aggregated or summarised the results and reported back using the generic scenario header 'project quality'.
Ensure adequate people and skills requirements for developing relevant risk scenarios.	Developing a manageable and relevant set of risk scenarios requires: • Expertise and experience, to not overlook relevant scenarios and not be drawn into highly unrealistic or irrelevant scenarios. While the avoidance of scenarios that are unrealistic or irrelevant is important in properly utilising limited resources, some attention should be paid to situations that are highly infrequent and unpredictable, but which could have a cataclysmic impact on the enterprise. • A thorough understanding of the environment. This includes the IT environment (e.g., infrastructure, applications, dependencies between applications, infrastructure components), the overall business environment, and an understanding of how and which IT environments support the business environment to understand the business impact. • The intervention and common views of all parties involved—senior management, which has the decision power; business management, which has the best view on business impact; IT, which has the understanding of what can go wrong with IT; and risk management, which can moderate and structure the debate amongst the other parties. • The process of developing scenarios usually benefits from a brainstorming/workshop approach, where a high-level assessment is usually required to reduce the number of scenarios to a manageable, but relevant and representative, number.
Use the risk scenario building process to obtain buy-in.	Scenario analysis is not just an analytical exercise involving 'risk analysts'. A significant additional benefit of scenario analysis is achieving organisational buy-in from enterprise entities and business lines, risk management, IT, finance, compliance and other parties. Gaining this buy-in is the reason why scenario analysis should be a carefully facilitated process.

Figure 1.29—Risk Scenario Technique Main Focus Areas *(cont.)*	
Focus/Issue	**Summary Guidance**
Involve first line of defence in the scenario building process.	In addition to co-ordinating with management, it is recommended that selected members of the staff who are familiar with the detailed operations be included in discussions, where appropriate. Staff whose daily work is in the detailed operations are often more familiar with vulnerabilities in technology and processes that can be exploited.
Do not focus only on rare and extreme scenarios.	When developing scenarios, one should not focus only on worst-case events because they rarely materialise, whereas less-severe incidents happen more often.
Deduce complex scenarios from simple scenarios by showing impact and dependencies.	Simple scenarios, once developed, should be further fine-tuned into more complex scenarios, showing cascading and/or coincidental impacts and reflecting dependencies. For example: • A scenario of having a major hardware failure can be combined with the scenario of failed DRP. • A scenario of major software failure can trigger database corruption and, in combination with poor data management backups, can lead to serious consequences, or at least consequences of a different magnitude than a software failure alone. • A scenario of a major external event can lead to a scenario of internal apathy.
Consider systemic and contagious risk.	Attention should be paid to systemic and/or contagious risk scenarios: • **Systemic**—Something happens with an important business partner, affecting a large group of enterprises within an area or industry. An example would be a nationwide air traffic control system that goes down for an extended period of time, e.g., six hours, affecting air traffic on a very large scale. • **Contagious**—Events that happen at several of the enterprise's business partners within a very short time frame. An example would be a clearinghouse that can be fully prepared for any sort of emergency by having very sophisticated disaster recovery measures in place, but when a catastrophe happens, finds that no transactions are sent by its providers and hence is temporarily out of business.
Use scenario building to increase awareness for risk detection.	Scenario development also helps to address the issue of detectability, moving away from a situation where an enterprise 'does not know what it does not know'. The collaborative approach for scenario development assists in identifying risk to which the enterprise, until then, would not have realised it was subject to (and hence would never have thought of putting in place any countermeasures). After the full set of risk items is identified during scenario generation, risk analysis assesses frequency and impact of the scenarios. Questions to be asked include: • Will the enterprise ever detect that the risk scenario has materialised? • Will the enterprise notice something has gone wrong so it can react appropriately? Generating scenarios and creatively thinking of what can go wrong will automatically raise and, hopefully, cause response to, the question of detectability. Detectability of scenarios includes two steps: visibility and recognition. The enterprise must be in a position that it can observe anything going wrong, and it needs the capability to recognise an observed event as something wrong.
Source: ISACA, *COBIT 5 for Risk*, USA, 2013, figure 37	

Business continuity teams are often working from the same type of scenarios as those developed for purposes of risk identification and assessment, and the risk practitioner may find it beneficial to partner with these teams as a means of avoiding unnecessary duplication of effort. Reviewing audit reports, interviewing end users or IT personnel, and observing the business processes or operations in action may also provide a useful basis for scenario development.

1.9 OWNERSHIP AND ACCOUNTABILITY

Risk requires ownership and accountability. After a risk has been identified, a manager or senior official in the organization must be identified as its owner. A risk owner is accountable for accepting risk based on the organizational risk appetite and should be someone with the budget, authority, and mandate to select the appropriate risk response based on analyses and guidance provided by the risk practitioner.

This accountability extends to approving controls when mitigation is the chosen risk response. Types of controls are examined in more detail in chapter 3 Risk Response and Mitigation. However, the concept is to create a direct link between risk and control, so that all risk is addressed through appropriate controls and all controls are justified by the risk that mandates their existence. This relationship is shown in **figure 1.30.**

Figure 1.30—Risk in Relation to Control

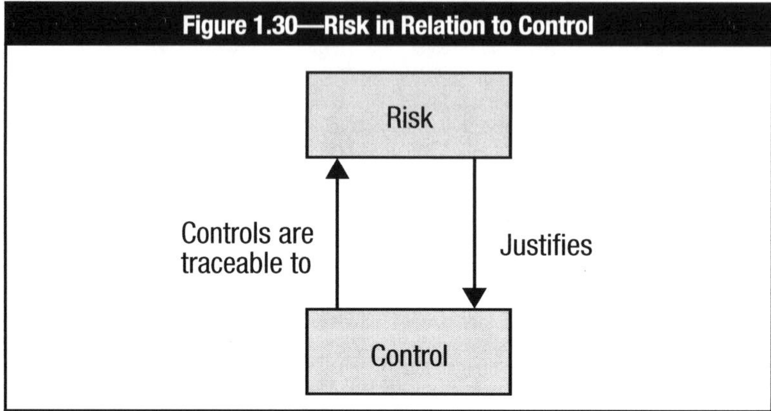

The owner of a risk also owns any controls associated with that risk and is accountable for monitoring their effectiveness. In some areas where there are regulations or laws that apply to risk, the risk owner may have to prepare standard reports on the status of risk, any incidents that may have occurred and the level of risk currently faced by the organization.

1.10 THE IT RISK REGISTER

Organizations maintain risk registers in order to consolidate all information about risk into one central repository. The risk register allows senior management and the managers of each department to obtain the status of the risk management process from a single source, which in turn makes it possible to better manage and report on risk and coordinate risk response activities. At any given time, the risk register should document the entire risk universe of the organization.

The risk register must be kept up to date in order to be effective. Risk practitioners should be familiar with the elements of the risk register and ensure that its content is periodically reviewed both to add new risk and remove risk that may no longer be relevant to the organization. An example of a risk register template is shown in **figure 1.31**.

Figure 1.31—Example of a Risk Register Template

Part I—Summary Data						
Risk statement						
Risk owner						
Date of last risk assessment						
Due date for update of risk assessment						
Risk category	☐ STRATEGIC (IT Benefit/Value Enablement)		☐ PROJECT DELIVERY (IT Programme and Project Delivery)		☐ OPERATIONAL (IT Operations and Service Delivery)	
Risk classification (copied from risk analysis results)	☐ LOW		☐ MEDIUM	☐ HIGH	☐ VERY HIGH	
Risk response	☐ ACCEPT		☐ TRANSFER	☐ MITIGATE	☐ AVOID	
Part II—Risk Description						
Title						
High-level scenario (from list of sample high-level scenarios)						
Detailed scenario description—Scenario components	**Actor**					
	Threat Type					
	Event					
	Asset/Resource					
	Timing					
Other scenario information						
Part III—Risk Analysis Results						
Frequency of scenario (number of times per year)	0	1	2	3	4	5
	N ≤ 0,01 ☐	0,01 < N ≤ 0,1 ☐	0,1 < N ≤ 1 ☐	1 < N ≤ 10 ☐	10 < N ≤ 100 ☐	100 < N ☐
Comments on frequency						
Impact of scenario on business	0	1	2	3	4	5
1. Productivity	**Revenue Loss Over One Year**					
Impact rating	I ≤ 0,1% ☐	0,1% < I ≤ 1% ☐	1% < I ≤ 3% ☐	3% < I ≤ 5% ☐	5% < I ≤ 10% ☐	10% < I ☐
Detailed description of impact						
2. Cost of response	**Expenses Associated With Managing the Loss Event**					
Impact rating	I ≤ 10k$ ☐	10K$ < I ≤ 100K$ ☐	100K$ < I ≤ 1M$ ☐	1M$% < I ≤ 10M$ ☐	10M$ < I ≤ 100M$ ☐	100M$ < I ☐
Detailed description of impact						
3. Competitive advantage	**Drop-in Customer Satisfaction Ratings**					
Impact rating	I ≤ 0,5 ☐	0,5 < I ≤ 1 ☐	1 < I ≤ 1,5 ☐	1,5 < I ≤ 2 ☐	2 < I ≤2,5 ☐	2,5 < I ☐
Detailed description of impact						
4. Legal	**Regulatory Compliance—Fines**					
Impact rating	None ☐	< 1M$ ☐	< 10M$ ☐	< 100M$ ☐	< 1B$ ☐	> 1B$ ☐
Detailed description of impact						
Overall impact rating (average of four impact ratings)						
Overall rating of risk (obtained by combining frequency and impact ratings on risk map)	☐ LOW		☐ MEDIUM	☐ HIGH	☐ VERY HIGH	

Figure 1.31—Example of a Risk Register Template *(cont.)*				
Part III—Risk Response				
Risk response for this risk	☐ ACCEPT	☐ TRANSFER	☐ MITIGATE	☐ AVOID
Justification				
Detailed description of response (NOT in case of ACCEPT)	**Response Action**		**Completed**	**Action Plan**
	1.		☐	☐
	2.		☐	☐
	3.		☐	☐
	4.		☐	☐
	5.		☐	☐
	6.		☐	☐
Overall status of risk action plan				
Major issues with risk action plan				
Overall status of completed responses				
Major issues with completed responses				
Part IV—Risk Indicators				
Key risk indicators for this risk	1. 2. 3. 4.			
Source: ISACA, *COBIT 5 for Risk*, USA, 2013, figure 62				

The purpose of a risk register is to consolidate risk data into one place and permit the tracking of risk. Entries in the risk register show the severity, source and potential impact of a risk, as well as identifying the risk owner and the current status and disposition of the risk. The risk register is the one document that contains all risk that has been detected by various parts and activities of the organization including risk identified in audits, vulnerability assessments, penetration tests, incident reports, process reviews, management input, risk scenario creation and security assessments. This allows management to refer to one place to gain insight into the outstanding risk issues, the status of risk mitigation efforts and the emergence of newly identified and documented risk. In some implementations, the risk register may be accompanied by a controls register, which includes information on testing and monitoring.

1.11 RISK AWARENESS

Awareness is a powerful tool in creating the culture, forming ethics and influencing the behavior of the members of an organization. The staff and operational teams of an organization are often the first to be aware of any problems or abnormal activities. Through awareness programs, it is possible to develop a team approach to risk management that enables every member of an organization to identify and report on risk and to work to defend systems and networks from attacks. Each member of the team can help identify vulnerabilities, suspicious activity and possible attacks. This may enable a faster response and better containment of a risk when an attack happens.

Risk awareness acknowledges that risk is an integral part of the business. This does not imply that all risk is to be avoided or eliminated, but rather that:

- Risk is well understood and known.
- IT risk issues are identifiable.
- The enterprise recognizes and uses the means to manage risk.

A risk awareness program creates an understanding of risk, risk factors and the various types of risk that an organization faces. An awareness program should be tailored to the needs of the individual groups within an organization and deliver content suitable for that group. A risk awareness program should not disclose vulnerabilities or ongoing investigations except where the problem has already been addressed. Using examples and descriptions of the types of attacks and compromises that other organizations have experienced can reinforce the need for diligence and caution when addressing risk.

Awareness education and training can serve to mitigate some of the biggest organizational risk and achieve the most cost-effective improvement in risk and security. This can generally be achieved by educating an organization's staff in required procedures and policy compliance, as well as ensuring that staff can identify and understand the risk that threatens the organization. It is critical that the training effectively communicate the risk and its potential impact in order for staff to understand the justification for what many see as inconvenient extra steps that risk mitigation and security controls often require.

The risk practitioner must also understand the organization's structure and culture as well as the types of communication that are most effective, to develop awareness and training programs that will be effective in the environment. Periodically changing risk awareness messages and the means of delivery will help maintain a higher level of risk awareness. Procedural controls can be complex, and it is essential to provide training as needed to ensure that staff understand the procedures and can correctly perform the required steps.

Awareness of information security policies, standards and procedures by all personnel is essential to achieving effective risk management. Employees cannot be expected to comply with policies or standards that they are not aware of or follow procedures they do not understand. The risk practitioner should advise for a standardized approach, such as short computer- or paper-based quizzes to gauge awareness levels. Periodic use of a standardized testing approach provides metrics for awareness trends and training effectiveness. Further training needs can be determined by a skills assessment or employing a testing approach. Indicators for additional training requirements can come from various sources such as tracking help desk activity, operational errors, security events and audits.

An awareness program for management should highlight the need for management to play a supervisory role in protecting systems and applications from attack. A manager has the responsibility to oversee the actions of his/her staff and direct compliance with the policies and practices of the organization.

Awareness training for senior management should highlight the liability, need for compliance, due care and due diligence and the need to create the tone and culture of the organization through policy and good practice. Senior management may need to be reminded that they are the ones who "own" the risk and bear the responsibility for determining risk acceptance levels.

1.12 SUMMARY

This chapter addresses the initial phase of a risk management effort. The intention of risk identification is to set out a clear path for the later chapters of risk assessment, risk response, and risk and control monitoring.

The core learning objectives of this chapter are to identify risk factor such as assets, threats, vulnerabilities, consequence and likelihood. These are documented and used in the next steps of the risk management process to assess the risk and ultimately drive the selection of appropriate controls or another suitable risk response.

ENDNOTES

1. ISACA, *COBIT 5 for Risk*, USA, 2013
2. *Ibid.*
3. *Ibid.*
4. Rittenburg, Larry; Frank Martens; *Enterprise Risk Management—Understanding and Communicating Risk Appetite*, Committee of Sponsoring Organizations of the Treadway Commission (COSO), USA, 2012
5. International Organization for Standardization/International Electrotechnical Commission (ISO/IEC); *ISO/IEC 27005:2011 Information Technology—Security Techniques—Information Security Risk Management*, Switzerland, 2011.
6. Verizon; *2015 Data Breach Investigations Report*, USA, 2015, *http://www.verizonenterprise.com/DBIR/2015/*
7. The Open Web Application Security Project (OWASP), *www.owasp.org*
8. ISACA, *Generating Value from Big Data Analytics*, USA, 2014, *www.isaca.org/Knowledge-Center/Research/Documents/Generating-Value-from-Big-Data-Analytics_whp_Eng_0114.pdf*
9. European Union, *Regulation EC No 45/2001: Data Protection by Community Institutions and Bodies, europa.eu/legislation_summaries/information_society/data_protection/l24222_en.htm*
10. ISO; *ISO 31000:2009 Risk Management—Principles and Guidelines*, Switzerland, 2009
11. *Op cit* ISACA, 2013
12. ISO; *IEC 31010:2009 Risk Management—Risk Assessment Techniques*, Switzerland, 2009
13. ISO; *ISO/IEC 27001:2013 Information Technology—Security Techniques—Information Security Management Systems—Requirements*, Switzerland, 2013
14. *Op cit* ISO/IEC 2011
15. National Institute of Standards and Technology (NIST); *NIST Special Publication 800-30 Revision 1: Guide for Conducting Risk Assessments*, USA, 2012
16. NIST; *NIST Special Publication 800-39: Managing Information Security Risk*, USA, 2011
17. *Op cit* ISACA, 2013

Chapter 2: IT Risk Assessment

Section One: Overview

Section Two: Contents

Section One: Overview

DOMAIN DEFINITION

Analyze and evaluate IT risk to determine the likelihood and impact on business objectives to enable risk-based decision making.

LEARNING OBJECTIVES

The objective of this domain is to ensure that the CRISC candidate has the knowledge necessary to:
- Identify and apply risk assessment techniques
- Analyze risk scenarios
- Identify current state of controls
- Assess gaps between current and desired states of the IT risk environment
- Communicate IT risk assessment results to relevant stakeholders.

CRISC EXAM REFERENCE

This domain represents 28 percent of the CRISC exam (approximately 42 questions).

TASK AND KNOWLEDGE STATEMENTS

TASKS

There are six tasks within this domain that a CRISC candidate must know how to perform. These relate to the IT risk assessment process.

T2.1 Analyze risk scenarios based on organizational criteria (e.g., organizational structure, policies, standards, technology, architecture, controls) to determine the likelihood and impact of an identified risk.
T2.2 Identify the current state of existing controls and evaluate their effectiveness for IT risk mitigation.
T2.3 Review the results of risk and control analysis to assess any gaps between current and desired states of the IT risk environment.
T2.4 Ensure that risk ownership is assigned at the appropriate level to establish clear lines of accountability.
T2.5 Communicate the results of risk assessments to senior management and appropriate stakeholders to enable risk-based decision making.
T2.6 Update the risk register with the results of the risk assessment.

KNOWLEDGE STATEMENTS

The CRISC candidate should be familiar with the task statements relevant to each domain in the CRISC job practice. The tasks are supported by 41 knowledge statements that delineate each of the areas in which the risk practitioner must have a good understanding in order to perform the tasks. Many knowledge statements support tasks that cross domains.

The CRISC candidate should have knowledge of:
1. Laws, regulations, standards and compliance requirements
2. Industry trends and emerging technologies
3. Enterprise systems architecture (e.g., platforms, networks, applications, databases and operating systems)
4. Business goals and objectives
5. Contractual requirements with customers and third-party service providers
6. Threats and vulnerabilities related to:
   6.1 Business processes and initiatives
   6.2 Third-party management
   6.3 Data management
   6.4 Hardware, software and appliances
   6.5 The system development life cycle (SDLC)

   6.6 Project and program management
   6.7 Business continuity and disaster recovery management (DRM)
   6.8 Management of IT operations
   6.9 Emerging technologies
7. Methods to identify risk
8. Risk scenario development tools and techniques
9. Risk identification and classification standards, and frameworks
10. Risk events/incident concepts (e.g., contributing conditions, lessons learned, loss result)
11. Elements of a risk register
12. Risk appetite and tolerance
13. Risk analysis methodologies (quantitative and qualitative)
14. Organizational structures
15. Organizational culture, ethics and behavior
16. Organizational assets (e.g., people, technology, data, trademarks, intellectual property) and business processes, including enterprise risk management (ERM)
17. Organizational policies and standards
18. Business process review tools and techniques
19. Analysis techniques (e.g., root cause, gap, cost-benefit, return on investment [ROI])
20. Capability assessment models and improvement techniques and strategies
21. Data analysis, validation and aggregation techniques (e.g., trend analysis, modeling)
22. Data collection and extraction tools and techniques
23. Principles of risk and control ownership
24. Characteristics of inherent and residual risk
25. Exception management practices
26. Risk assessment standards, frameworks and techniques
27. Risk response options (i.e., accept, mitigate, avoid, transfer) and criteria for selection
28. Information security concepts and principles, including confidentiality, integrity and availability of information
29. Systems control design and implementation, including testing methodologies and practices
30. The impact of emerging technologies on design and implementation of controls
31. Requirements, principles, and practices for educating and training on risk and control activities
32. Key risk indicators (KRIs)
33. Risk monitoring standards and frameworks
34. Risk monitoring tools and techniques
35. Risk reporting tools and techniques
36. IT risk management best practices
37. Key performance indicator (KPIs)
38. Control types, standards, and frameworks
39. Control monitoring and reporting tools and techniques
40. Control assessment types (e.g., self-assessments, audits, vulnerability assessments, penetration tests, third-party assurance)
41. Control activities, objectives, practices and metrics related to:
   41.1 Business processes
   41.2 Information security, including technology certification and accreditation practices
   41.3 Third-party management, including service delivery
   41.4 Data management
   41.5 The system development life cycle (SDLC)
   41.6 Project and program management
   41.7 Business continuity and disaster recovery management (DRM)
   41.8 IT operations management
   41.9 The information systems architecture (e.g., platforms, networks, applications, databases and operating systems)

SELF-ASSESSMENT QUESTIONS

CRISC self-assessment questions support the content in this manual and provide an understanding of the type and structure of questions that have typically appeared on the exam. Questions are written in a multiple-choice format and designed for one best answer. Each question has a stem (question) and four options (answer choices). The stem may be written in the form of a question or an incomplete statement. In some instances, a scenario or a description problem may also be included. These questions normally include a description of a situation and require the candidate to answer two or more questions based on the information provided. Many times a question will require the candidate to choose the **MOST** likely or **BEST** answer among the options provided.

In each case, the candidate must read the question carefully, eliminate known incorrect answers and then make the best choice possible. Knowing the format in which questions are asked, and how to study and gain knowledge of what will be tested, will help the candidate correctly answer the questions.

2-1 The **MOST** significant drawback of using quantitative risk analysis instead of qualitative risk analysis is the:

A. lower objectivity.
B. greater reliance on expertise.
C. less management buy-in.
D. higher cost.

2-2 Risk scenarios are analyzed to determine:

A. strength of controls.
B. likelihood and impact.
C. current risk profile.
D. scenario root cause.

2-3 The risk to an information system that supports a critical business process is owned by:

A. the IT director.
B. senior management.
C. the risk management department.
D. the system users.

2-4 The **PRIMARY** reason risk assessments should be repeated at regular intervals is:

A. omissions in earlier assessments can be addressed.
B. periodic assessments allow various methodologies.
C. business threats are constantly changing.
D. they help raise risk awareness among staff.

2-5 Which of the following choices **BEST** assists a risk practitioner in measuring the existing level of development of risk management processes against their desired state?

A. A capability maturity model (CMM)
B. Risk management audit reports
C. A balanced scorecard (BSC)
D. Enterprise security architecture

2-6 Which of the following choices **BEST** helps identify information systems control deficiencies?

A. Gap analysis
B. The current IT risk profile
C. The IT controls framework
D. Countermeasure analysis

2-7 Deriving the likelihood and impact of risk scenarios through statistical methods is **MOST LIKELY** to be associated with which type of risk analysis?

A. risk scenario
B. qualitative
C. quantitative
D. semiquantitative

2-8 Which of the following reviews is **BEST** suited for the review of IT risk analysis results before the results are sent to management for approval and use in decision making?

A. An internal audit review
B. A peer review
C. A compliance review
D. A risk policy review

ANSWERS TO SELF-ASSESSMENT QUESTIONS

Correct answers are shown in **bold**.

2-1 A. Neither of the two risk analysis methods is fully objective. While the qualitative method subjectively assigns high, medium and low frequency and impact categories to a specific risk, subjectivity within the quantitative method is often expressed in mathematical "weights."
B. To be effective, both processes require personnel who have a good understanding of the business.
C. Quantitative analysis generally has a better buy-in than qualitative analysis to the point where it can cause overreliance on the results.
D. Quantitative risk analysis is generally more complex and, therefore, more costly than qualitative risk analysis.

2-2 A. The strength of controls is determined after the controls are in place to ensure they are adequate in addressing the risk.
B. Risk scenarios are descriptions of events that can lead to a business impact and are evaluated to determine the likelihood and impact should the event occur.
C. The current risk profile is the identification of risk currently of concern by the organization.
D. The risk scenario process is used to identify plausible scenarios and from there determine likelihood and impact. Determining a root cause is not a part of the risk scenario process.

2-3 A. The IT director manages the IT systems on behalf of the business owners.
B. Senior management is responsible for the acceptance and mitigation of all risk.
C. The risk management department determines and reports on level of risk but does not own the risk.
D. The system users are responsible for utilizing the system properly and following procedures, but they do not own the risk.

2-4 A. Performing risk assessments on a periodic basis can find omissions in earlier assessments, but this is not the primary reason for conducting regular reassessments.
B. Organizations strive to improve their risk management process to more quickly and accurately assess and address risk, and this may involve changing the methodology. However, it is not the primary reason for conducting regular assessments.
C. As business objectives and methods change, the nature and relevance of threats also change. This is the primary reason to conduct periodic risk assessments.
D. Risk assessments are conducted on a periodic basis to address new threats and changes in the business. Creating more risk awareness is a minor benefit of conducting periodic risk assessments.

2-5 **A. The capability maturity model (CMM) grades processes on a scale of 0 to 5, based on their maturity. It is commonly used by entities to measure their existing state and then to determine the desired one.**
B. Risk management audit reports offer a limited view of the current state of risk management.
C. A balanced scorecard (BSC) enables management to measure the implementation of strategy and assists in its translation into action.
D. Enterprise security architecture explains the security architecture of an entity in terms of business strategy, objectives, relationships, risk, constraints and enablers and provides a business-driven and business-focused view of security architecture.

2-6 **A. Controls are deployed to achieve the desired control objectives based on risk assessments and business requirements. The gap between desired control objectives and actual IS control design and operational effectiveness identifies IS control deficiencies.**
B. Without knowing the gap between desired state and current state, one cannot identify the control deficiencies.
C. The IT controls framework is a generic document with no information such as desired state of IS controls and current state of the enterprise; therefore, it will not help in identifying IS control deficiencies.
D. Countermeasure analysis only helps in identifying deficiencies in countermeasures, not in the full set of primary controls.

2-7 A. A risk scenario analysis might include any of several risk analysis methods, including quantitative, semi-quantitative and qualitative; it is not reflective of a particular approach.
B. A qualitative risk analysis uses experiential and subjective measures to estimate the likelihood and impact of adverse events according to ranges; these might include low, medium and high ratings for both likelihood and impact.
C. The essence of quantitative risk assessment is to derive the likelihood and impact of risk scenarios based on statistical methods and data.
D. Semi-quantitative analysis typically applies to a wider, numerically delineated range of values to a qualitative rating mechanism—for example, assigning values from 0 to 100. The assignment remains qualitative, and it is not associated with statistical analysis.

2-8 A. An internal audit review is not best suited for the review of IT risk analysis results. Internal auditing is an independent, objective assurance and consulting activity designed to add value and improve an enterprise's operations. It helps an organization accomplish its objectives by bringing a systematic, disciplined approach to evaluate and improve the effectiveness of risk management, control and governance processes.
B. It is effective, efficient and good practice to perform a peer review of IT risk analysis results before sending them to management.
C. A compliance review is not best suited for the review of IT risk analysis results. Compliance reviews measure the conformance with a specific, measurable standard.
D. A review of the risk policy will change the contents and methods of the risk analysis eventually, but this is not a way of reviewing IT risk analysis results before sending them to management.

Note: For more self-assessment questions, you may also want to obtain a copy of the *CRISC™ Review Questions, Answers & Explanations Manual 4th Edition* or the Database, which each consist of 500 multiple-choice study questions, answers and explanations.

SUGGESTED RESOURCES FOR FURTHER STUDY

In addition to the resources cited throughout this manual, the following resources are suggested for further study in this domain (publications in **bold** are stocked in the ISACA Bookstore):

Committee of Sponsoring Organizations of the Treadway Commission (COSO); *Enterprise Risk Management for Cloud Computing*, USA, 2012

International Organization for Standardization (ISO); *International Electrotechnical Commission (IEC) 31010:2009: Risk management—Risk assessment techniques*, Switzerland, 2009

ISO/IEC; *ISO/IEC 27005:2008: Information technology—Security techniques—Information security risk management*, Switzerland, 2008

ISACA, COBIT® 5, USA, 2012, *www.isaca.org/cobit*

ISACA, *COBIT® 5 for Risk*, USA, 2013, *www.isaca.org/cobit*

ISACA, *The Risk IT Framework*, USA, 2009

ISACA, *The Risk IT Practitioner Guide*, USA, 2009

Kissel, Richard; Kevin Stine; Matthew Scholl; Hart Rossman; Jim Fahlsing; Jessica Gulick; *NIST Special Publication 800-64 Revision 2: System Considerations in the System Development Life Cycle*, National Institute of Standards and Technology (NIST), USA, 2008

National Institute of Standards and Technology (NIST); *NIST Special Publication 800-30 Revision 1: Guide for Conducting Risk Assessments*, USA, 2012

Panda, Parthajit, "The OCTAVE Approach to Information Security Risk Assessment," *ISACA Journal*, Volume 4, 2009

Section Two: Content

2.0 OVERVIEW

After risk is identified and documented in the risk register, the next step in the risk management life cycle shown in **figure 2.1** is to assess the IT risk level. IT risk is a subset of enterprise risk, and the risk faced by an IT system is most often measured by the impact of an IT-related problem on the business services that the IT system supports. Calculations or assessments of IT impact must consider the dependencies of the other systems, departments, business partners and users on the affected IT system.

Figure 2.1—The IT Risk Management Life Cycle

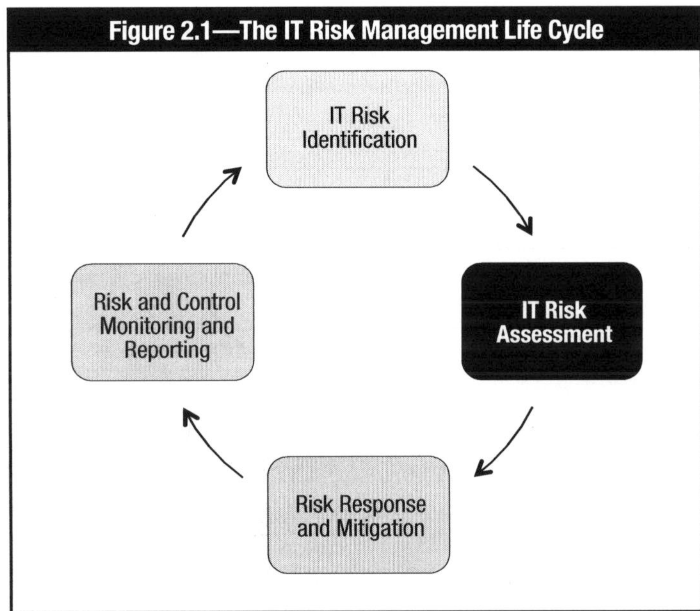

Risk assessment is a process used to identify and evaluate risk and its potential effects, which includes evaluation of the:

- Critical functions necessary for an enterprise to continue business operations
- Risk associated with each of the critical functions
- Controls in place to reduce exposure and their cost
- Prioritization of the risk on the basis of their likelihood and potential impact
- Relationship between the risk and the enterprise risk appetite and tolerance

The risk assessment process generates the information used to respond to risk in an appropriate and cost-effective manner.

2.1 RISK ASSESSMENT TECHNIQUES

There are several techniques used in risk assessment, including those listed in **figure 2.2**.

Figure 2.2—Risk Assessment Techniques
Bayesian statistics and Bayes nets
Bow tie analysis
Brainstorming/Structured or semistructured interviews
Business impact analysis (BIA)
Cause and consequence analysis
Cause-and-effect analysis
Checklists
Delphi method

Figure 2.2—Risk Assessment Techniques *(cont.)*
Environmental risk assessment
Event tree analysis
Fault tree analysis
Hazard analysis and critical control points (HACCP)
Hazard and operability study (HAZOP)
Human reliability analysis (HRA)
Layers of protection analysis (LOPA)
Markov analysis
Monte Carlo simulation
Preliminary hazard analysis
Reliability-centered maintenance
Root cause analysis (pre-mortems)
Scenario analysis
Sneak circuit analysis
Structured "what if" technique (SWIFT)
Adapted from *IEC 31010:2009 Risk management—Risk assessment techniques*, Switzerland, 2009

A consistent risk assessment technique should be used whenever the goal is to produce results that can be compared over time. However, risk practitioners need not rely exclusively on one (or even several) techniques. Each approach has certain advantages and possible weaknesses, and the risk practitioner should choose a technique appropriate for the circumstances of the assessment.

2.1.1 Bayesian Analysis

A Bayesian analysis is a method of statistical inference that uses prior distribution data to determine the probability of a result. This technique relies on the prior distribution data to be accurate in order to be effective and to produce accurate results.

2.1.2 Bow Tie Analysis

A bow tie analysis provides a diagram to communicate risk assessment results by displaying links between possible causes, controls and consequences. The cause of the event is depicted in the middle of the diagram (the "knot" of the bow tie) and triggers, controls, mitigation strategies and consequences branch off of the "knot."

2.1.3 Brainstorming/Structured Interview

The structured interview and brainstorming model gathers a large group of types of potential risk or ideas to be ranked by a team. The initial interview or brainstorming may be completed using prompts or interviews with an individual or small group. For more information on interviewing techniques, see section 1.7.1 Conducting Interviews.

2.1.4 Business Impact Analysis

Business impact analysis (BIA) is a process to determine the impact of losing the support of any resource. In addition to identifying initial impact, a comprehensive BIA seeks to establish the escalation of loss over time. The goal of BIA is to provide reliable data on the basis of which senior management can make the appropriate decision.

2.1.5 Cause and Consequence Analysis

A cause and consequence analysis combines techniques of a fault tree analysis and an event tree analysis and allows for time delays to be considered.

2.1.6 Cause-and-effect Analysis

A cause-and-effect analysis looks at the factors that contributed to a certain effect and groups the causes into categories (using brainstorming), which are then displayed using a diagram, typically a tree structure or a fishbone diagram.

2.1.7 Checklists

A checklist is a list of potential or typical threats or other considerations that should be of interest to the organization, whose items can be checked off one at a time as they are completed. The risk practitioner may use previously developed lists, codes or standards to assess the risk using this method.

2.1.8 Delphi Method

The Delphi method leverages expert opinion received using two or more rounds of questionnaires. After each round of questioning, the results are summarized and communicated to the experts by a facilitator. This collaborative technique is often used to build a consensus among experts.

2.1.9 Event Tree Analysis

An event tree analysis is a forward-looking, bottom-up model that uses inductive reasoning to assess the probability of different events resulting in possible outcomes.

2.1.10 Fault Tree Analysis

A fault tree analysis starts with an event and examines possible means for the event to occur (top-down) and displays these results in a logical tree diagram. This diagram can be used to generate ways to reduce or eliminate potential causes of the event.

2.1.11 Hazard Analysis and Critical Control Points (HACCP)

Originally developed for the food safety industry, HACCP is a system for proactively preventing risk and assuring quality, reliability and safety of processes. The system monitors specific characteristics, which should fall within defined limits.

2.1.12 Hazard and Operability Studies (HAZOP)

HAZOP is a structured means of identifying and evaluating potential risk by looking at possible deviations from existing processes.

2.1.13 Human Reliability Analysis (HRA)

HRA examines the effect of human error on systems and their performance.

2.1.14 Layers of Protection Analysis (LOPA)

LOPA is a semi-quantitative risk analysis technique that uses aspects of HAZOP data to determine risk associated with risk events. It also looks at controls and their effectiveness.

2.1.15 Markov Analysis

A Markov analysis is used to analyze systems that can exist in multiple states. The Markov model assumes that future events are independent of past events.

2.1.16 Monte-Carlo Analysis

IEC 31010:2009 describes Monte Carlo simulation in the following manner:[1]

> *Monte Carlo simulation is used to establish the aggregate variation in a system resulting from variations in the system, for a number of inputs, where each input has a defined distribution and the inputs are related to the output via defined relationships. The analysis can be used for a specific model where the interactions of the various inputs can be mathematically defined. The inputs can be based upon a variety of distribution types according to the nature of the uncertainty they are intended to represent. For risk assessment, triangular distributions or beta distributions are commonly used.*

2.1.17 Preliminary Hazard Analysis

Preliminary hazard analysis looks at what threats or hazards may harm an organization's activities, facilities or systems. The result is a list of potential risk.

2.1.18 Reliability-centered Maintenance

Reliability-centered maintenance analyzes the functions and potential failures of a specific asset, particularly a physical asset such as equipment.

2.1.19 Root Cause Analysis

Root cause analysis is a process of diagnosis to establish the origins of events, which can be used for learning from consequences, typically from errors and problems. It is discussed in further detail in section 2.6.3 Root Cause Analysis.

2.1.20 Scenario Analysis

Scenario analysis examines possible future scenarios that were identified during risk identification, looking for risk associated with the scenario should it occur. Section 2.3 discusses this technique in more detail.

2.1.21 Sneak Circuit Analysis

A sneak circuit analysis is used to identify design errors or sneak conditions such as latent hardware, software or integrated conditions that are often undetected by system tests and may result in improper operations, loss of availability, program delays or injury to personnel.

2.1.22 Structured "What If" Technique (SWIFT)

A structured "what if" technique uses structured brainstorming to identify risk, typically within a facilitated workshop. It uses prompts and guide words and is typically used with another risk analysis and evaluation technique.

2.2 ANALYZING RISK SCENARIOS

During risk identification, risk scenarios are developed and used to identify and describe potential risk events. These scenarios are useful to communicate with the business and gather input data required to understand the potential or probable impact of the risk event if it were to occur.

The impact of a risk event is hard to calculate with any degree of accuracy because there are many factors that affect the outcome of an event. If the event is detected quickly and appropriate measures are taken to contain the incident, then the impact may be minimized and the recovery process may be fairly rapid. However, if the organization is unable to detect the incident promptly, the same incident could cause severe damage and result in much higher recovery costs. Some of the factors that can affect the calculation of risk assessment are discussed in the following sections.

2.2.1 Organizational Structure and Culture

The structure and culture of the organization are contributing factors in risk prevention, risk detection and risk response. A mature organization has policies and procedures and an effective reporting and notification structure in place to detect, notify and escalate a situation effectively. For instance, an organization that does not have a mature incident response capability will often react to incidents in an unpredictable, *ad hoc* reactive manner and can expect to experience inconsistent results.

The risk management function should have an enterprise mandate that allows risk practitioners to review and provide input into all business processes, participate in incident management activities and be responsible for reviewing incidents for lessons learned and to improve incident response planning, detection and recovery. Lessons learned in one department or with regards to one system or application may be applicable in protecting other departments, systems, or applications from the same problems, so collaboration and sharing of information is an important part of using risk scenarios. If the culture of the organization is to hide problems rather than communicate or address them—or to only use an adverse situation to point blame—then the ability of the risk practitioner to effectively contribute to the protection of the organization and assist in the investigation of an incident may be severely impaired.

2.2.2 Policies, Standards and Procedures

Policies provide direction regarding acceptable and unacceptable behaviors and actions to the organization. Standards and procedures support the requirements defined in the policies set by the organization.

Policies

Policies empower risk management, audit and security staff. The policies of the organization should clearly state the position of senior management toward the protection of information, which will allow the development of procedures, standards and baselines that reflect management priorities. Executive sponsorship also provides a mandate that all departments comply with policy requirements.

Organizations often have several layers of policy in order to allow delegation of authority. A high-level policy may be issued by senior management as a way to address the objectives of the organization's mission and vision statement. This overarching policy does not have a technical focus in order to prevent it from becoming outdated when technology changes. High-level policy may direct compliance with laws and good practices, and is likely to state the goal of managing risk through protecting the organization's assets, including the information and information systems that support business operations.

High-level policies are instrumental in determining the approach of the organization toward risk management and acceptable levels of risk. Without policies in place, the risk practitioner may not be able to gain access to key personnel, be left out of strategic planning sessions and be ignored when performing investigations. Below the high-level policies are technical and functional policies that include specifics regarding technology use, such as remote access, acceptable use and passwords. These policies are subject to change as technology changes and new systems are developed.

The risk practitioner should identify the presence or lack of policies and work to determine whether the policies are enforced. If policies are not developed and communicated, the organization has no means of enforcing standards of behavior, which increases the risk that the behavior that occurs may be inappropriate. A lack of enforcement may also lead to circumvention of controls or increased liability because the organization has admitted the need for a policy but does not follow its own rules. The risk practitioner should assess the risk associated with the policy framework of the organization and provide recommendations as necessary. Where policies are out of date, lack enforcement or are incomplete, the risk practitioner should clearly document the vulnerability and associated risk that this poses to the organization.

Standards

A standard is a mandatory requirement, code of practice or specification approved by a recognized external standards organization, such as the International Organization for Standardization (ISO). Standards are implemented to comply with the requirements and direction of policy to limit risk and support efficient business operations.

Many organizations realize that the value of standards is the authority and proven value that they provide. A standard mandates the way in which personnel in an organization must comply with recognized practices or specifications and can help ensure a consistent approach to meeting risk requirements across an organization. An organization may base its practices and operations on external standards such as an ISO standard, or it may develop its own standards, such as requiring all staff to use the same product, operating system or desktop. The use of a standard facilitates support and maintenance, provides better cost control, and provides authority for the practices and procedures of the organization because a standard requires the implementation of certain practices.

Procedures

Procedures are more granular than standards and support their implementation. A procedure is a document containing a detailed description of the steps necessary to perform specific operations in conformance with applicable standards. Procedures are defined as part of processes and created in order to define the ways in which the processes should be carried out. Procedures are invaluable as a means of implementing the intent of policy, because they define the tasks that people actually carry out. By defining actions in consistent and measureable ways, the organization can greatly increase the probability that an operation is conducted according to good practice. Therefore, the risk practitioner can be assured that any abnormal operations will be detected.

A lack of standards and procedures makes it difficult to carry out activities in a systematic manner and may result in undependable, inconsistent operations and elevated risk. The risk practitioner should also distinguish between the existence of published procedures and their actual use. It is common for procedures to be followed only for a short time, after which point experienced staff begin to work from memory. This practice should be discouraged in any environment in which precision is important, such as shutdown procedures for power plants or industrial machinery or complex monetary transactions.

Exception Management

Having policies, standards and procedures in place is an essential part of operating secure systems and attaining a secure state; however, there may be cases in which an exception to policy, standard or procedure is necessary. If exceptions are undocumented and uncontrolled, the level of risk is unknown and may represent a hidden vulnerability. Therefore, exceptions should only be allowed through a documented, formal process that requires approval of the exception by a senior manager. The risk practitioner must ensure that an exception management process is in place and is being followed and that exceptions are removed when they are no longer needed.

2.2.3 Technology

The age and condition of technology used by an organization present a substantial risk factor. Many organizations use outdated technologies that are difficult to obtain, support and maintain. The technology in use may have been acquired through multiple projects and/or mergers. The technology may consist of products from a varied mix of vendors, languages, configurations and vintages, or it may run on different operating systems, hardware, network architectures and databases. A large organization may have systems based on mainframes, client-servers, virtual environments and the cloud all at once. The complexity that this represents is a source of risk.

In general, organizations that retain legacy systems do so because there is a compelling business need. It is good practice to have a simple, easily managed and controlled environment with standard products and technologies, but this may be nearly impossible to attain in a large organization. Many systems may continue to be used long after their anticipated life span simply because it is not cost-effective to replace them.

Some of the considerations that affect risk assessment related to technology include:
- Age of equipment
- Expertise available for maintenance
- Variety of vendors/suppliers (are they still in business?)
- Documentation of systems
- Availability of replacement parts
- Ability to test systems or equipment
- Operating environment and user expertise
- Ability to patch/mitigate vulnerabilities

The risk practitioner should work with business process owners and technical staff to determine what legacy systems are in use. It is also important verify both that their continued use represents a compelling business need and that the risk associated with such use has been documented and accepted by senior management. In many cases, legacy systems persist out of need, but managers are unaware of the extent to which the organization incurs risk from their persistence. Under such circumstances, the role of the risk practitioner is to advise the risk owner of ways in which the assessed risk might be addressed in order to bring it into alignment with the organizational risk appetite (or at least to tolerable levels).

2.2.4 Architecture

A key factor in the maturation of the processes and practices of an organization is the development of enterprise approaches to risk management, architecture and business continuity. Having an enterprise approach does not mean that different divisions or business areas are forced to accept a cookie-cutter model. These areas should tailor their approach through how requirements are met, not by varying what is required. Addressing these functions at the enterprise level promotes consistency, repeatability, compliance and accountability. It also improves the visibility afforded to senior management with regards to the practices and strategy of the organization.

Relatively few organizations have a mature IT architecture. In many organizations, business processes rely on systems built as part of individual projects or initiatives, and each system is an independent entity with little in common with other systems. The lack of an enterprise architecture results in ownership gaps between systems and unclear areas of responsibility for incident or configuration management.

As the complexity of an architecture increases, it invariably becomes more challenging for the organization that owns it to secure it and ensure compliance with security standards, regulations and good practices. The risk practitioner may find that systems and networks in the organization are not mapped out on a network diagram, and some systems and remote-access capabilities tied to legacy applications may not be documented at all. Controls may overlap or conflict with one another, and there may be methods to bypass controls or single points of failure. Under such circumstances, the ability of the organization to measure risk or identify vulnerabilities is severely impacted. The risk practitioner should consider adoption of a coherent, structured enterprise architecture a high priority.

2.2.5 Controls

When assessing risk, the risk practitioner must take into consideration the current control environment. Controls are implemented to reduce or maintain risk at acceptable levels; however, controls may be poorly maintained, unsuitable for the risk that they are meant to control or incorrectly configured. Controls must be regularly reviewed to determine and ensure their effectiveness.

In addition to evaluating the effectiveness of each control for its intended purpose, the risk practitioner should verify the correct balance between technical, managerial (administrative) and physical or operational control types. Implementation of a technical control such as a firewall, requires training for the staff who manages or operates it, correct procedures for its configuration, assignment of responsibilities for its monitoring , and schedules for regular testing. If these coinciding controls are not in place, stakeholders may develop a false sense of security, resulting in unidentified vulnerabilities, an ineffective use of resources, and greater risk than anticipated or intended.

Controls are generally categorized as belonging to one of six categories, described in **figure 2.3**.

Figure 2.3—Control Categories

Category	Description
Preventive	Inhibit attempts to violate security policy. Encryption, user authentication, and vault-construction doors are examples of preventive controls.
Deterrent	Provide warnings that may dissuade threat agents from attempting compromise. Warning banners on login screens and rewards for the arrest of hackers are examples of deterrent controls.
Directive	Mandate behavior by specifying what actions are and are not permitted, which may also have a deterrent effect. A policy is an example of directive control.
Detective	Provide warning of violations or attempted violations of security policy. Audit trails, intrusion detection systems (IDSs) and checksums are examples of detective controls.
Corrective	Remediate errors, omissions, unauthorized uses and intrusions when detected. Data backups, error correction and automated failover are examples of corrective controls.
Compensating	An alternate form of a control that corrects a deficiency or weakness in the control structure of the enterprise. This may be considered when an entity cannot meet a stated requirement due to legitimate technical or business constraints but can create a comparably acceptable level of risk by other means. Placing unsecured systems on isolated network segments with strong perimeter security and adding third-party challenge-response mechanisms to devices that do not support individual login accounts are examples of compensating controls.

The interaction between the control types is shown in **figure 2.4**.

Figure 2.4—Control Category Interdependencies

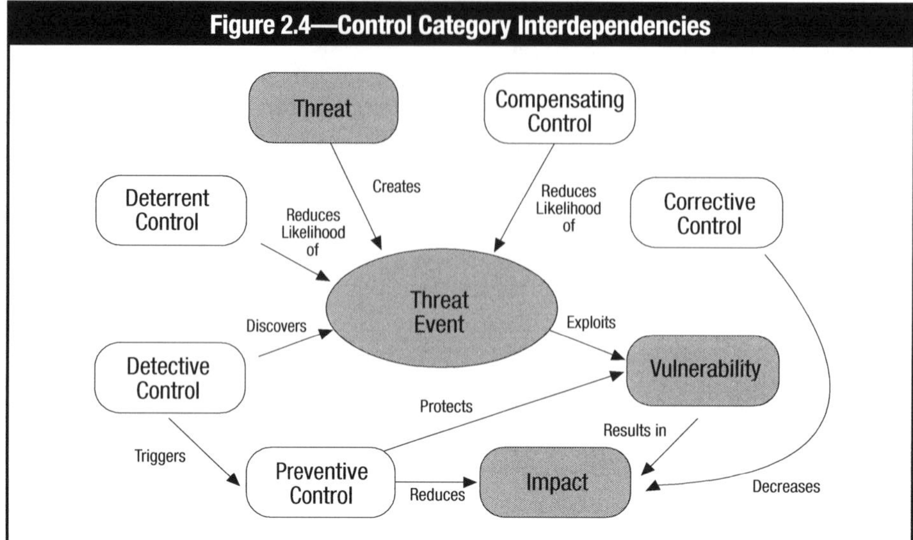

Assessing the control environment provides the risk practitioner with an opportunity to evaluate the risk culture and effectiveness of the current risk management program, which can be used to determine both the level of risk currently facing the organization and the seriousness of that risk. In general, risk is more serious when any of the following are true:

- Controls are inadequate.
- The wrong controls are being used.
- Controls are ignored or bypassed.
- Controls are poorly maintained.
- Logs or control data are not reviewed.
- Controls are not tested.
- Changes to the configuration of controls are not managed.
- Controls can be physically accessed and altered.
- Duties are inadequately segregated. Individuals are able to perform two or more of the following:
  - Approve changes
  - Make changes
  - Monitor changes
  - Analyze changes
  - Report on changes

The selection of controls is addressed in chapter 3 Risk Response and Mitigation; monitoring is addressed in chapter 4 Risk and Control Monitoring and Reporting.

2.3 CURRENT STATE OF CONTROLS

The risk practitioner determines the current state of IT risk using the reports generated by controls themselves and the results of control testing activities and incident management programs. A current-state assessment refers to the condition of the program at a point in time. An IT risk assessment should be regarded as accurate only at the time the risk state is measured. However, regular reviews of IT risk may be used to determine the state of IT risk on a regular basis and with scheduled reporting to management.

For the purposes of IT risk assessment, the risk practitioner must review current risk levels and contrast the current level with the desired state or level of acceptable risk. A gap or disparity between the current and desired state should be investigated to determine its reason as well as to identify workable solutions that may address the disparity. Some tools that may be used by the risk practitioner to determine the current state of IT risk include:

- Audits
- Business continuity plans (BCPs)
- Capability maturity models

- Control tests conducted by the control owner or custodian
- Incident reports
- IT operations and management evaluation
- Enterprise architecture assessment
- Logs
- Media reports of new vulnerabilities and threats
- Observation
- Self-assessments
- Third-party assurance
- User feedback
- Vendor reports
- Vulnerability assessments and penetration tests

2.3.1 Audits

An audit is a formal inspection and verification to check whether a standard or set of guidelines is being followed, records are accurate or efficiency and effectiveness targets are being met. Recommendations from audits are often related to the improvement of managerial, technical and operational controls. Audits tend to be taken seriously by senior management because of their strong reliance on evidence and use of published, rigorous methodology. The risk practitioner may encounter reluctance from risk owners when asking to bring in auditors because there is little room to argue with an audit finding, which some managers see as negative appraisals of their ability. Where use of auditors is permitted, their assistance can be invaluable.

2.3.2 Business Continuity Plans

All systems will fail from time to time. The purpose of business continuity planning is to enable a business to continue critical services in the event of a disruption, up to and including the ability to survive a disastrous interruption. Rigorous planning and commitment of resources is necessary to adequately plan for such an event, and the first step in preparing a new BCP is to identify the business processes of strategic importance, meaning those key processes that are responsible for both the permanent growth of the business and for the fulfillment of the business goals. BCP development should be supported by a formal policy that states the organization's overall target for recovery and empowers those people involved in developing, testing and maintaining the plans.

The risk practitioner is typically not responsible for BCP development, but he or she is likely to work closely with the BCP teams because BCPs are controls that reduce the risk of losing critical processes to acceptable levels. In particular, the risk practitioner may be involved in assessing the adequacy of BCPs, which depends on the levels of risk that they are meant to address. As with any other asset, the risk associated with a particular process depends on the magnitude of impact to the business should it be interrupted and the probability of an interruption. Thus, the result of the risk assessment should be the identification of the following:

- The human resources, data, infrastructure elements and other resources (including those provided by third parties) that support the key processes
- A list of potential vulnerabilities (the dangers or threats to the organization)
- The estimated probability of the occurrence of these threats
- The efficiency and effectiveness of existing risk mitigation controls (risk countermeasures)

BCPs may be established at the enterprise level, by each department or for each process When there is an overarching plan, it is usually divided into multiple annexes that relate to the division and process-level activities that must be carried out to continue operations. It is highly desirable to have a single integrated plan to ensure that:

- There is proper coordination among various plan components.
- Resources committed are used in the most effective way and there is reasonable confidence that, through its application, the organization will survive a disruption.

The goal of a BCP is to provide a reduced but sufficient level of functionality in the business operations immediately after encountering an interruption and while recovery is taking place, so the plan should address all functions and assets required to continue as a viable organization. A BCP includes the continuity procedures determined by the

organization as necessary to survive and minimize the consequences of business interruption. The BCP also lists the people needed to carry out the continuity procedures. The recovery of critical business processes may be through an alternate process, including:
- A manual process or outsourced support
- Having sufficient inventory on hand to support operations
- Using facilities available at another office or location
- Displacing less critical work with more critical functions

The core source of data used in business continuity planning is the BIA, which identifies the critical time lines for services and products associated with value creation. The BIA also establishes the recovery point objective (RPO) and recovery time objective (RTO) for a process, which respectively defines how much data can be lost in recovery and how quickly it must be accomplished. Because the BCP is based on the BIA, the risk practitioner must review the process used to determine the BIA to validate that it is accurate and considers all of the risk factors. In organizations that have adopted a comprehensive business continuity management system (BCMS), the risk practitioner may benefit from having direct access to BCPs for review and reference in the context of risk management.

Disaster Recovery

Disaster recovery refers to the reestablishment of business and IT services following a disaster or incident within a predefined schedule and budget; the term is commonly associated with IT rather than business, but disaster recovery can be considered a relative of overall business continuity. Time frames for recovery are based on the cost of recovery and length of time that management is willing to accept. The risk practitioner should review the disaster recovery plan (DRP) to ensure that it is up to date and reflects risk scenarios and business priorities. A comprehensive DRP includes specific information on hardware and software requirements for restoration, which systems and applications should be restored in what order, how to accomplish the restorations under multiple scenarios, and how many user logins are required in what time frames. The plan should also be rigorously tested, which can uncover any vulnerabilities and provide the experience needed for team members to enact the plan effectively.

2.3.3 Capability Maturity Models

The risk practitioner may find it useful to compare the state of the organization's risk management program to an established model of capability maturity, which contains the essential elements of effective processes for one or more disciplines. It also describes an evolutionary improvement path from *ad hoc*, immature processes to disciplined, mature processes with improved quality and effectiveness.

A mature organization has defined, reliable processes that it follows consistently and continuously seeks to improve. As a result, an organization with a capable and mature risk management process is much more likely to prevent incidents, detect incidents sooner and recover rapidly from incidents. The organization should seek to implement well-structured risk management procedures across all of its departments and regions and ensure that every business process and systems development project follows the organization's core risk management principles, policies, procedures and standards. The risk management framework should reflect the risk appetite of the organization because it is the core of the organization's risk culture.

Key elements used to measure IT risk management capability include:
- Support of senior management
- Regular communication between stakeholders
- Existence of policies, procedures and standards
- Availability of a current BIA
- Logging and monitoring of system activity
- Regular review of logs
- Scheduled risk assessments and review
- Testing of BCPs and DRPs
- Training of staff
- Involvement of risk principles and personnel in IT projects
- Gathering feedback from users and stakeholders
- Validating the risk appetite and risk acceptance levels
- Time to detect/resolve a security incident

Improvements in both the efficiencies and effectiveness of the risk management capability are based on the consistent application of policies and procedures. If the procedures are followed on a consistent basis, then the efficiency and effectiveness of the risk management practices will improve.

2.3.4 Control Tests

Testing a control includes testing both the technical and nontechnical aspects of the control, such as the rules governing the operation of the control, the procedures used in monitoring and operating the control and the proficiency of the staff responsible for the operation of the control. Controls are selected to address one or more specific types of risk. Controls may not necessarily be designed, implemented or maintained in a manner that provides the desired level of protection. The risk practitioner should test the performance of controls and advise the risk owner of any instance in which a control is being bypassed, disabled or rendered ineffective through changes to the operating environment, the configuration of the control or a previously unidentified threat.

2.3.5 Incident Reports

Incident management starts with the preparation and planning that build an incident response plan (IRP). The organization should prevent incidents where possible but also have the controls in place to detect and respond to an incident when it occurs. The sooner an incident can be detected, the easier it is to contain the incident and minimize the damage or impact caused by the incident. The response team must have training on the IRP and the skills necessary to investigate the incident. Incident response teams (IRTs) generally consist of employees assigned specific roles during various types of incidents, but external resources may be needed to perform forensic investigations.

The primary focus of incident management is to get the organization's affected systems and operations back into normal service as quickly as possible. This can impact the ability to secure evidence associated with the incident because evidence may be lost while returning to service. A detailed IRP and regular testing that involves the business will increase awareness and ensure the appropriate time is granted for evidence collection and analysis while still meeting the business RTOs and RPOs. If the business does not understand the need to collect evidence or the IRT does not know how to collect and preserve evidence, the organization is at a greater risk to be liable for the event, and the attacker has a greater chance to avoid prosecution.

Each incident must be examined to extract the lessons learned to improve the prevention, detection and recovery from future incidents of a similar nature. A thorough review of an incident can identify weak controls, poor detection, inappropriate or ineffective response and lack of training of staff. The risk practitioner may learn about threats, vulnerabilities and impact from the review of incident reports.

2.3.6 IT Operations and Management Evaluation

IT operations staff is usually among the first to be aware of problems relating to systems and applications. Interviewing operations staff and reviewing logs and trouble tickets can provide insight into an unmitigated or recurring problem within a system, which should be investigated to determine whether they pose unaccepted risk to the organization.

IT operations typically include multiple functions related to risk management and quality assurance, including policies and procedures for the management of incidents, configurations, problems, and releases of new systems or software. The risk practitioner should work with IT staff to understand all of these functions and how they are carried out. Weaknesses in any of the IT functions may create opportunities for threat agents to exploit systems or processes.

Many elements of IT risk are a result of how an IT department is run and managed. Where an attitude of complacency reigns, risk ignorance can result in breaches of security. Security problems commonly seen within the IT department include sharing high-level accounts and passwords and the lack of control over changes to production systems. The pressure to complete projects on time or meet demands from the business can result in shortcuts that pose significant unaccepted risk to the business.

Data Management

The risk practitioner should review and assess the data ownership and management process of the organization, including the protection of data from improper disclosure, modification or deletion. A lack of clear ownership of the data may be a significant risk. The rules for access should be maintained and enforced, and all staff should be aware of the risk associated with improper data management (such as data leakage) and their requirement to comply with the policies and standards of the organization. The risk practitioner must pay considerable attention when assessing the risk associated with data management to ensure compliance with data management policies and procedures.

IT management must play an active role in mitigating risk and supporting risk management activities. Above all, IT managers must abide by their own rules. The example set by IT management is crucial to developing a positive risk culture and promoting the integration of risk management principles into IT systems and projects. The risk practitioner should be observant of daily practices in the IT department and bring undocumented risk to the attention of accountable managers.

2.3.7 Enterprise Architecture Assessment

Enterprise architecture (EA) focuses on producing a view of the current state of IT, establishing a vision for a future state and generating a strategy to get there. This view of IT should demonstrate links between IT and organizational objectives and produce a view of current risk and controls and answer or enable answering of the four AREs, shown in **figure 2.5**.

Figure 2.5— "Four Ares"

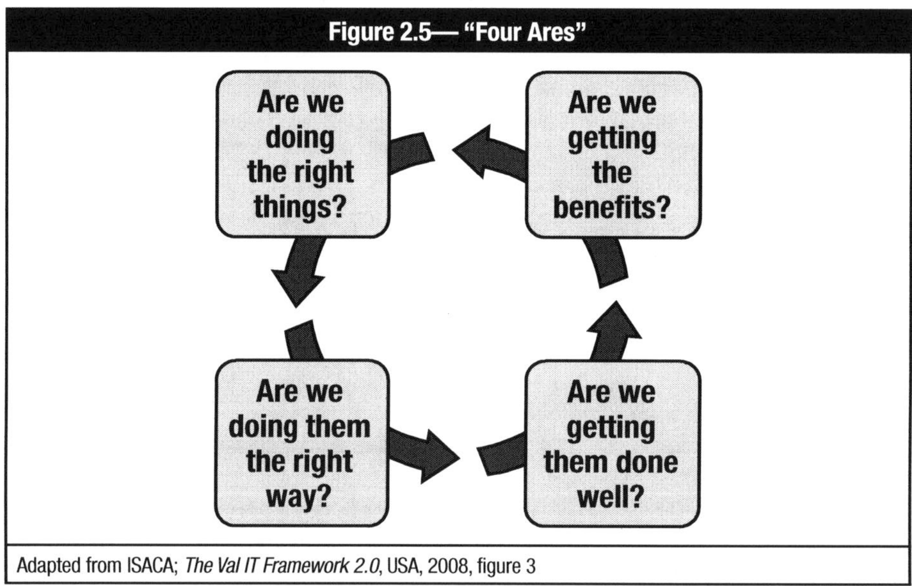

Adapted from ISACA; *The Val IT Framework 2.0*, USA, 2008, figure 3

In any knowledge-based organization or an organization dependent on information systems, EA is a key component of the control environment. The risk associated with an IT system is a combination of the risk associated with each element that makes up an IT system: a risk in any one location can affect the security of all other areas, and a vulnerability on one system can affect the security of all other systems.

The risk practitioner should inquire as to the existence of an enterprise architecture and assess the EA to determine its maturity. Significant effort, resources and time are required to develop an EA, and it is not uncommon for the risk practitioner to discover that the EA is at a low maturity state. Risk management requires a complete and thorough assessment of the risk to each element of a system. In environments in which an EA is either immature or absent, the risk practitioner must place greater emphasis on the use of technology-specific assessments as means of building a piecemeal vision of current IT risk.

2.3.8 Logs

Logs are one of the most valuable tools to monitor controls and detect risk, but they are often underutilized. Common challenges relating to the effective use of logs include:

- Having too much data
- Difficulty in searching for relevant information
- Improper configuration (e.g., may not be enabled or contain appropriate data)
- Modification or deletion of data before it is read (e.g., too little storage space)

A log should contain a record of all important events that occur on a system, such as:

- Changes to permissions
- System startup or shutdown
- Login or logout
- Changes to data
- Errors or violations
- Job failures

Log reviews can identify risk-relevant events such as compliance violations, suspicious behavior, errors, probes or scans and abnormal activity. A failure to review the logs can result in the organization not being aware of an ongoing attack. Logs should also be preserved for forensic analysis if needed at a later time. The risk practitioner may find it useful to employ analysis tools when reviewing logs in order to filter pertinent data.

Ensuring proper segregation of duties (SoD) is particularly important when it comes to log files. The ability to change system configurations should be segregated from the ability to review, modify or delete logs in order to ensure that the organization can exercise proper oversight of administrative functions.

2.3.9 Media Reports

Media sources—such as newspapers, television, and articles—can be useful sources of information about threats or incidents when considered with a healthy degree of professional skepticism. The risk practitioner should alert those with authority when a media communication could impact the organization or its employees, customers or business partners and may provide advice on how to respond to a threat mentioned in the media that could affect the organization or a product or service used by the company's internal or external stakeholders. All stakeholders need to be educated on potential risk associated to the news feed by the appropriate department in the organization.

2.3.10 Observation

Watching a process as an independent observer may highlight issues that are unable to be seen as clearly when in the middle of daily operations and help to identify situations in which documented processes are not being followed. Evaluating exercises and real-world activities from the perspective of an observer is particularly valuable for the risk practitioner, who should first consider what is actually happening and ensure that it is reflected in workflows; later, these workflows can be compared to stated procedures.

2.3.11 Self-assessments

Many organizations encourage direct manager involvement in monitoring risk and control effectiveness within their areas of responsibility. Local managers typically have insight into the behavior of their staff and daily operations, so they are well-suited to evaluate compliance with procedures, recurring problems, risk trends and vulnerabilities. Properly conducted self-assessment can provide the risk team and senior management with regular reporting of risk and assurance of control effectiveness and may also reduce the need for more intense audits or system reauthorization.

2.3.12 Third-party Assurance

Third-party performance of reviews, audits, compliance verification and analysis of the organization, its processes, incidents, logs and threat environment can be a valuable source of information about risk, control effectiveness, vulnerabilities and compliance. The organization may also benefit from the expertise, objectivity and credibility of a third-party organization, whose assurance that the organization is managing risk adequately and is compliant with standards and good practices may be seen as more credible than the results of an assessment conducted purely within the organization.

Third-party Management

Where organizations rely on third parties to carry out key aspects of their business functions, the need for third-party assurance takes on a different connotation. If a third party is going to perform work for the organization, nondisclosure agreements (NDAs) are necessary to protect the intellectual property of the organization from being disclosed to unauthorized personnel. This is especially relevant when a penetration test or other assessment has been performed that could be used to perpetrate an attack on the organization. Outsourcing services are also subject to disasters, even when they boast of highly redundant data center environments replicated across continents.

The relationship between an outsourcing provider and the organization is defined by contract. Legal requirements written into contracts should address the jurisdiction for any complaints or disagreements with the provider, and regulations in the host country regarding disclosure to law enforcement and data transmission and storage across borders. The risk practitioner should ensure that security and regulatory requirements are addressed in all agreements with suppliers and service providers. Service level agreements (SLAs) are also commonly included in outsourcing contracts, but the risk practitioner should make certain that risk owners do not misinterpret them. SLAs provide monetary remedies. Outsourcing does not terminate the ownership or liability of a risk owner for data or performance; for instance, the organization remains accountable for the security of information that it stores with a third-party host. Standards of due care and/or due diligence may be warranted in certain situations or regulated industries.

The risk practitioner should take steps to ensure any provider of outsourced services used by the organization is compliant with good practices, whether through external audits such as Statement on Standards for Attestation Engagements Number 16 (SSAE 16) or US Health Information Technology for Economic and Clinical Health (HITECH) reports or having the right to conduct an audit on behalf of the organization. Where possible, reviewing an outsourcing provider's security policies and procedures may provide an additional level of assurance. This can be difficult to arrange unless it is written into the original agreement, so the risk practitioner should be aware and involved in all discussions regarding the establishment of third-party agreements to ensure that executives make decisions with a clear understanding of the associated risk.

While the term "cloud" is relatively new, remote data hosting is a service that has been around for many years. The cloud refers to the use of the Internet (usually drawn as a cloud in network diagrams) to transmit data to a remote location for storage or processing and includes software as a service (SaaS), storage on demand, infrastructure as a service (IaaS) and platform as a service (PaaS). The risk issues associated with cloud computing are the same as for any other outsourcing initiative.

2.3.13 User Feedback

System users are in an excellent position to know the system and its shortcomings, including potential vulnerabilities. Interacting with users and learning their issues and concerns regarding the system may indicate where security controls could be circumvented for convenience or improved efficiency. For instance, if users are regularly exporting data from a system in order to create or customize reports or perform analysis using basic office applications, the risk practitioner should verify whether such practices are formally recognized by procedure or have been adopted on an ad hoc basis, in which case they may present unaccepted risk.

2.3.14 Vendor Reports

There are many sources for information on current threats and vulnerabilities, new types of malware or emerging attack methods. Many government-sponsored computer emergency response teams (CERTs) and security vendors provide free or subscription-based reports and analysis, and these can be valuable sources of information for the risk practitioner.

2.3.15 Vulnerability Assessments and Penetration Tests

The risk practitioner may find it useful to review vulnerability assessment results and penetration testing reports as a means of estimating the level of IT risk in the organization. These technical evaluations tend to be very detailed and can be shared and discussed directly with IT staff. Presenting them to senior management often requires a degree of translation to the language of business. One advantage of these types of testing is that their results are likely to be granted a level of deference by senior management similar to an audit, but its conduct may be more easily accepted by IT management because it falls within the realm of IT.

Vulnerability assessment and penetration testing are discussed in more depth in section 1.5.3 and section 4.5.2.

2.4 CHANGES IN THE RISK ENVIRONMENT

The risk environment changes as organizations and technologies change and threat agents focus on new areas to attack. Systems and applications that were once thought to be secure are often found to be vulnerable at a later time. This requires the ongoing monitoring and evaluation of risk (see chapter 4 Risk and Control Monitoring and Reporting).

Substantial changes in the risk environment may arise as a result of changes in technology or in business practices. The risk practitioner should ensure that new and emerging risk is identified, assessed and reported and that the organization is aware of and watching for emerging threats and vulnerabilities. This includes monitoring vendor alerts, reports from CERTs and media stories. When a new issue appears in the media, the risk practitioner should work with the business and system owner to perform a threat analysis and determine if and how the organization should respond.

2.4.1 Emerging Technologies

New technologies are coming on the market continuously. The pressure to implement new technology is often influenced by inflated expectations of its utility and maturity and a focus on product functionality without attention paid to security. The desire to deploy new technologies for competitive advantage may cause the organization to lose sight of the business risk involved unless the risk practitioner is diligent and makes a compelling case for robust testing and assessment. Foresight and forecasting practices and techniques are risk management activities that have a longer-term outlook.

Emerging technologies often provide indicators years in advance of their potential. It is the job of the risk practitioner to consider the potential risk and controls for the application of these technologies that may present value to the organization. A well-managed change control process ensures that new technologies are not implemented until the security team has been able to validate the security impact of the change and enable appropriate controls. However, it is not good enough to wait until a technology has been implemented before actually reviewing and securing it. In many cases, if the IT department is unwilling to review and integrate new technologies, the business will find ways to implement them regardless, putting the organization in a vulnerable position.

The risk practitioner should evaluate and assess the approach of the organization to accepting new technologies and the attitude of the security team and IT operations teams toward reviewing and securing the new technologies as they become available. The risk practitioner is considerably more likely to get the attention and support of senior management by leading the effort to demonstrate how new technologies can be safely incorporated into the enterprise than by adopting an obstructionist posture.

2.4.2 Industry Trends

Marketing staff is concerned with changes in trends, and a failure to see a changing trend can result in an organization losing a substantial percentage of its market share within a very short time period. This also applies to IT. A failure by the IT department to adapt to or support a new business model may result in substantial loss to the organization. For example, when some telecommunications providers switched to per second instead of per minute billing, it forced all of the other telecommunications providers to change their systems quickly or lose a large percentage of their customers.

A risk practitioner should assess the maturity of the IT department and the organization as a whole toward monitoring and adapting to new market trends. A lack of flexibility or poor communication between the business units and IT could be a risk factor.

2.5 PROJECT AND PROGRAM MANAGEMENT

There is considerable risk associated with the management of projects as well as programs. A program is defined as a structured grouping of interdependent projects that is both necessary and sufficient to achieve a desired business outcome and create value.

IT projects fail for various reasons, including:
- Unclear or changing requirements
- Scope creep (additional requirements)

- Lack of budget
- Lack of skilled resources
- Problems with technology
- Delays in delivery of supporting elements/equipment
- Unrealistic time lines (push to market)
- Lack of progress reporting

When a project is at risk, it is important to identify the root cause of the problem and take steps to address it as soon as possible. Lack of good project management can lead to:
- Loss of business
- Loss of competitive advantage
- Low morale among staff members
- Inefficient processes
- Lack of testing of new systems or changes to existing systems
- Impact on other business operations
- Failure to meet SLAs or contractual requirements
- Failure to comply with laws and regulations

When a project is at risk, the entire program may also be at risk because the program may rely on each project completing on time and on schedule.

2.5.1 The System Development Life Cycle

One method used to support systems development and IT projects is the system development life cycle (SDLC). The SDLC describes project management as a series of phases that provides structure and auditability to a project. An example of the SDLC is seen in **figure 2.6**.

Figure 2.6—Characteristics of the SDLC Phases

SDLC Phase	Phase Characteristics	Support from Risk Management Activities
Phase 1—Initiation	The need for an IT system is expressed and the purpose and scope of the IT system is documented.	Identified risk is used to support the development of the system requirements, including security requirements and a security concept of operations (strategy).
Phase 2—Development or Acquisition	The IT system is designed, purchased, programmed, developed, or otherwise constructed.	Risk identified during this phase can be used to support the security analyses of the IT system that may lead to architecture and design trade-offs during system development.
Phase 3—Implementation	The system security features should be configured, enabled, tested and verified.	The risk management process supports implementation against its requirements and within its modeled operational environment. Decisions regarding risk identified must be made prior to system operation.
Phase 4—Operation or Maintenance	The system performs its functions. Typically the system will undergo periodic updates or changes to hardware and software; the system may also be altered in less obvious ways due to changes to organizational processes, policies and procedures.	Risk management activities are performed for periodic system reauthorization (or reaccreditation or whenever major changes are made to an IT system in its operational, production environment (e.g., new systems interfaces).
Phase 5—Disposal	This phase may involve the disposition of information, hardware and software. Activities may include moving, archiving, discarding or destroying information and sanitizing the hardware and software.	Risk management activities are performed for system components that will be disposed of or replaced to ensure that the hardware and software are properly disposed of, that residual data are appropriately handled, and that system migration is conducted in a secure and systematic manner.
Source: ISACA, *CISM Review Manual 14th Edition*, USA, 2015, figure 2.31		

Regardless of the methodology used for project management, the core principles are the same. Proper oversight, clear requirements, user involvement, communications between team members and users, and regular review of project progress are all critical to project success. The risk practitioner will often find projects that are at risk of failure and then must identify the cause of the failure, recommend a solution and report the risk to management.

System development projects fail for many reasons, including:

- The requirements may change:
  - Scope creep
  - New business priorities
  - Poorly understood initial requirements
- Resources (trained staff or budget) may not be available.
- Dependencies (e.g., suppliers, outsourcers, communications providers, technology) may not deliver on time.
- Technology may not work as anticipated.
- Project complexity is underestimated.
- Resources are poorly managed:
  - Lack of leadership
  - Lack of accountability
  - Lack of oversight
- Symptoms of failure are not recognized.
- There is a lack of coordination with suppliers.

Regardless of the reason for the problems the project is experiencing, the risk to the organization is considerable for project failure. Project failure may result in:

- Indirect financial loss (loss of competitive advantage)
- Direct financial loss (violations of contract or SLAs)
- Inability to adjust to a changing operational environment
- Damage to reputation
- Dissatisfaction among the project team (decreased morale)

The identification of a project that is at risk of failure is important so that corrective action can be implemented as soon as possible. The challenge of identifying project risk is the problem of obtaining accurate data on project status and being able to identify the root cause of the project problem. Proper and engaged oversight and monitoring of the project and validation of progress reports may be required to prevent inaccurate project status reporting. An objective risk assessment is often instrumental in determining true project status. A project that is struggling to remain on time and on budget with the necessary resources may result in critical steps (such as testing) being skipped at the end of the project to get back on schedule. This can lead to production failures, inaccurate reporting and data processing, or missing functionality as the project is implemented.

The risk practitioner should verify that project design, development and testing did not solely focus on the business functions of the project. Security requirements built into the project time line should have been validated, and it should be tested to show that these reduce the risk associated with confidentiality, integrity and availability. The risk management process should be incorporated into each step of the SDLC, and security requirements should be integrated into the project life cycle. Early planning and awareness can result in substantial cost and time saving through proper risk management planning. Security discussions should be performed as part of (not separate from) the development project to ensure solid understanding among project personnel of business decisions and their risk implications to the overall development project.[2]

Key tasks to perform during the SDLC to expressly document the risk associated with the development of a new program include:

- Security categorization of the proposed system—What are its confidentiality, integrity and availability requirements?
- BIA—What impact would an outage have on critical business processes?
- Privacy impact assessment—What laws or regulations apply? What sensitive data are processed, stored or transmitted by this system?

- Use of a secure information systems development process—Is there security training for all development staff and a secure environment? Are there secure code practices?
- Awareness of vulnerabilities with selected technology or operational environment

The project must also provide the ability to audit and review the applications or systems after implemented to ensure proper operation of the system.

2.6 RISK AND CONTROL ANALYSIS

Risk assessment seeks to identify the level of risk relative to a level that is acceptable to the organization. The risk practitioner makes this assessment by comparing the current state of risk against the desired state, taking into account the effectiveness of controls that may already be in place. The desired state of IT risk is closely linked to the risk acceptance level set by senior management (and certainly should not exceed it). Therefore, the risk practitioner must first establish the risk appetite of the organization. The key goal of the risk assessment process is to identify all cases in which the current level of risk exceeds the acceptable risk level so that this information can be used as the basis for risk response.

2.6.1 Data Analysis

The risk practitioner has a wealth of data sources available, including network devices, application logs and audit reports. Having data is valuable, but events may be hidden by sheer volume. Therefore, the incorrect analysis of data may lead to an erroneous conclusion.

The challenges of data analysis start with the completeness and trustworthiness of the data:
- Are all of the data available?
- Have any of the data been altered or changed?
- Are the data in the correct format?
- Are the data based on measuring important factors?

Some of the approaches to conducting data analysis are listed in **figure 2.7.**

Figure 2.7—Methods for Uncovering Less Obvious Risk Factors

Method	Description
Cause-and-effect analysis	A predictive or diagnostic analytical tool that is used to: • Explore the root causes or factors that contribute to positive or negative effects or outcomes • Identify potential risk **Note:** A typical form is the Ishikawa diagram, also known as the fishbone diagram.
Fault tree analysis	A technique that: • Provides a systematic description of the combination of possible occurrences in a system, which can result in an undesirable outcome (top-level event) • Combines hardware failures and human failures
A fault tree is constructed by:	• Relating the sequences of events that, individually or in combination, could lead to the top-level event • Deducing the preconditions for the: – Top-level event – Next levels of events, until the basic causes are identified (elements of a "perfect storm" [unlikely simultaneous occurrence of multiple events that cause an extraordinary incident]) **Note:** The most serious outcome is selected as the top-level event.
Sensitivity analysis	A quantitative risk analysis technique that: • Helps to determine which risk factors potentially have the most impact • Examines the extent to which the uncertainty of each element affects the target object when all other uncertain elements are held at their baseline values **Note:** The typical display of results is in the form of a tornado diagram.

When analyzing data to determine risk levels, the risk practitioner should be attentive to the trends of events in the data sources. Trends may reveal an emerging threat, but detection of threats that are not already known is possible only when regular analysis has already established a reliable baseline of what constitutes normal behavior.

2.6.2 Threat and Misuse Case Modeling

Chapter 1 IT Risk Identification discusses the mindset and approach of the adversary and the extent to which motivation and skill level of the attacker are important in determining the real existence of a threat. Threat modeling examines the nature of the threat and potential threat scenarios, which is a valuable tool in the hands of the risk practitioner and can greatly assist in risk assessment.

Threat modeling is done by mapping the potential methods, approaches, steps and techniques used by an adversary to perpetrate an attack. The threat agent will often try different tools, probe for vulnerabilities, and try both technical and nontechnical approaches while seeking to compromise a system. The risk practitioner must think of as many of these methods and approaches as possible so that adequate controls can be designed to meet the possible threats. Threat modeling helps the risk practitioner, systems designers, developers and operators to build systems with attention to defensive controls, built-in security features and proper placement within a strategy of overlapping defenses (defense in depth) that can deter or prevent system failures.

Threat modeling and misuse case modeling differ from use case modeling, which examines how a system will function to deliver value to its users. Misuse case modeling looks at all the possible errors, mistakes, and intentional deviations from expected user behavior that a system may endure. Understanding misuse cases helps to ensure that a system is built with resiliency and the ability to handle errors and misuse.

Threat modeling is similar and examines the ways a system can be attacked and used for a purpose for which the system was never intended. An example of this is the "ping of death" attack. The Internet Control Message Protocol (ICMP) was designed as a tool for system and network administrators to test network connectivity, but by altering the size of an ICMP packet, an attacker is able to build an attack tool that could disable target system. Similar threats arising from intentional misuse have used the Network Time Protocol (NTP) and domain name system (DNS) services as attack platforms.

2.6.3 Root Cause Analysis

Root cause analysis is a process of diagnosis to establish the origins of events so that these can be used for learning from consequences (typically errors and problems). The actions that an organization takes in response to risk are often based on the lessons learned from previous events. A prudent manager examines the root cause of an incident to discover the conditions and factors that led to the event, rather than reacting to the symptoms of the problem. These underlying conditions, which comprise the root (core) of a problem, are the conditions against which action must be taken in order to prevent the problem from recurring. For example, if a review of a business process finds that users are not compliant with the procedures and policies in place, the risk practitioner should examine why the users are not compliant before recommending enforcement of the procedure. It may be that the procedure is outdated, flawed or unworkable when aligned with the objectives of the organization, in which case it is the procedure and not the behavior that should be corrected. Root cause analysis examines the reasons why a problem exists or a breach has occurred and seeks to identify and resolve these underlying issues.

One implementation of root cause analysis is a pre-mortem, a facilitated workshop where the group is told to pretend that the project has failed and discuss why it has failed. When correctly facilitated, the simple but crucial cognitive impact of answering the question in this fashion (instead of why it *might* fail) can produce insightful, collaborative and valuable perspectives on risk.

In many cases, a risk event may be the result of coinciding events—several issues that act in combination to create what appears to be a single result. The risk practitioner can use root cause analysis as a means of identifying coinciding events, which cannot be traced to a single common cause. In this case, the solution is more than just a single response activity or implementation of a single control.

2.6.4 Gap Analysis

By documenting the desired state or condition of risk that management wants to reach and then carefully analyzing and evaluating the current condition of the organization, the risk practitioner can identify the existence of a risk gap and the scope of actions that may be needed to close the gap. Gap analysis can be used in iteration to devise deliverables, projects and milestones. This process is illustrated in **figure 2.8**.

Figure 2.8—Availability and Gap Analysis

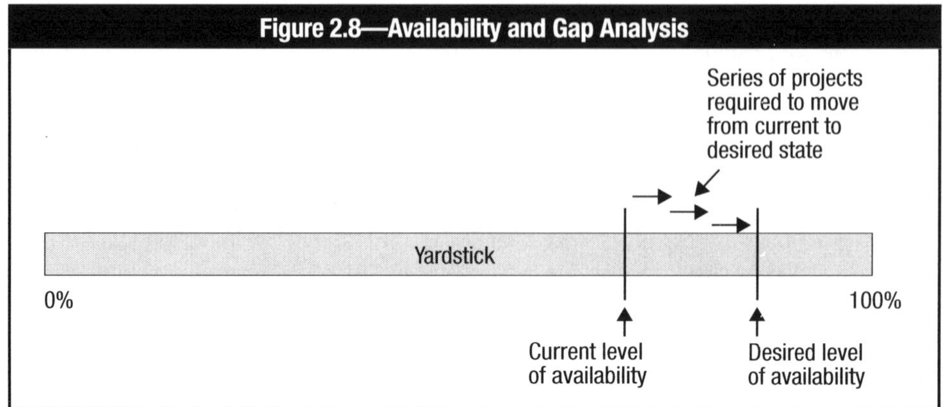

Using the gap analysis method to plan projects, along with milestones, can help the organization initiate and execute projects in an orderly and logical manner that takes into account their dependencies among each other.

To measure the current state, the risk practitioner may find it beneficial to identify one or more key indicators that relate to goals or performance (i.e., key goal indicators [KGIs] or key performance indicators [KPIs]), which may be internal to the organization to relate to identified standards, frameworks or good practices. Desired states are typically linked to risk (i.e., key risk indicators [KRIs]). KPIs and KRIs are discussed in section 4.1 Key Risk Indicators.

2.6.5 Predicting Risk

Many times, a risk is the result of preexisting or predisposing conditions that have been present for an extended period prior to their exploitation. Risk events may also cascade from one another, where one seemingly minor problem or vulnerability serves as a trigger that leads to a serious problem such as a prolonged outage. The risk practitioner should consider this when assessing risk and look for instances in which factors that are innocuous individually may collectively serve as trigger events.

Previous incidents, audit reports and failures affecting both internal and external parties provide the risk practitioner with valuable insight. Each event offers a wealth of information on what may have gone wrong and how to better protect systems, detect incidents and respond more effectively in the future. The greatest threat to the incident response process is a failure to learn from past events and repeat the same errors in the future. Investigations of past events may also validate the calculation of risk impact, but the risk practitioner is cautioned to remember that no two incidents are identical and that the impact of a future event may differ substantially from a past event with similar characteristics.

2.7 RISK ANALYSIS METHODOLOGIES

Risk analysis is a complex and important process that is often needed to provide the data necessary for effective risk response. Risk may be analyzed on either a quantitative or qualitative basis or in a way that reflects some combination of the two (semiquantitative).

2.7.1 Quantitative Risk Assessment

Quantitative risk assessment is based on numerical calculations, such as monetary values. Its reliance on numbers makes it precise. It is particularly suitable for cost-benefit analysis because risk that can be mapped to monetary values can easily and directly be compared to the costs of various risk responses. Unfortunately, the cost associated with a particular risk can be very difficult to quantify, especially if it includes subjective elements such as reputation and employee morale. Many events used in quantitative assessment are also unpredictable. The challenges of calculating likelihood are discussed in section 1.3.6 Likelihood/Probability.

Where quantitative risk assessment is desirable, the risk practitioner may seek to approximate probability using empirical or historical data to calculate an average likelihood over an entire population. Applying such results to individual cases is almost completely unreliable, but it may be suitable in certain cases. For instance, although mean time between failure (MTBF) is not an accurate predictor of failure for an individual appliance, an organization that has thousands of identical appliances in use can reasonably expect some number of them to begin failing at or around the time indicated by the MTBF for their make and model.

Quantitative risk assessment becomes progressively more useful as likelihood approaches certainty, which is a scenario that can be applied to hurricanes, blizzards and other sorts of weather events that are prevalent in certain geographical regions. In such cases, the need to account for the probability of an event is eliminated, because the event itself can be assumed. The risk practitioner is then left to account only for the impact to the organization of the associated consequences, which may be measured very precisely in terms of direct cost associated with damage to property and loss of revenue.

The longer that an adverse situation persists, however, the more difficult it becomes for the risk practitioner to estimate the impact. It also tends to be more difficult to predict the impact of events that have a wide scope, because capabilities that an organization assumes will be available may be stretched beyond capacity by the demands imposed upon them by other organizations in the affected area. One example of this phenomenon is that fire departments have a limited number of ambulances, which may result in delayed access to medical care.

The cost of risk used in quantitative risk assessment is often calculated on an annual basis in order to align the process with the natural cycle for calculating budgets. The risk practitioner may find it beneficial to compare various events that occur at different frequencies in a similar manner using a common denominator.

2.7.2 Qualitative Risk Assessment

Qualitative risk assessment assigns values on a comparative or ordinal basis (such as high, medium and low or a scale of 1 to 10). The assignment of qualitative values relies heavily on experience and expert knowledge, but such reliance does not make qualitative risk assessment less effective than quantitative risk assessment. Many types of risk are extremely difficult to calculate in purely numeric terms, and controlling for the unknown variables in a quantitative process can require so many assumptions as to render the apparently precise results relatively unhelpful. In contrast, the relative values offered by a qualitative process can typically be used to order response actions in terms of perceived importance. When the perception is based on a broad enough sample of stakeholders, the resulting course of action is likely to be acceptable to those stakeholders.

Qualitative risk assessments are usually based on scenarios or descriptions of situations that either have occurred or may occur. The intention of these scenarios is to elicit feedback from multiple stakeholders (e.g., departments, customers, management). By communicating with all affected shareholders, the risk practitioner is able to eliminate unlikely assumptions and determine a level of risk that reflects a reasonable consensus of feedback from all potentially affected groups.

Scenarios under qualitative risk assessment may be based on threats, vulnerabilities, assets/impact or some combination of these factors. A threat-based scenario examines a risk event on the basis of threat agents and seeks to identify potential methods of attack. From this information, it can be inferred which vulnerabilities will be exploited by the attacker along with the intent and skill of the attacker and the potential damage to the assets affected. This method is especially beneficial when examining the emergence of new threats and determining the risk related to advanced persistent threats (APTs), whose capabilities and available time make waiting for suitable vulnerabilities more feasible than is true of conventional attackers.

A vulnerability-based approach takes the opposite approach: it examines the organization's known vulnerabilities and then attempts to anticipate threats that could exploit those vulnerabilities, projecting from these the consequences and magnitude of impact. Vulnerability-based scenarios are especially valuable after completing vulnerability assessments, and their findings can be further assessed by carrying out penetration testing to determine whether the anticipated threats are feasible in the context of the organization and its systems.

An asset/impact approach is based on the identification of critical and sensitive assets and the potential ways that these could be damaged—typically by attacking their confidentiality, integrity and availability.

The results of a qualitative risk assessment are typically presented in the form of a table that compares the likelihood of a risk event with its impact to the organization, where the confluence of the two factors generates the relative level of the risk. For instance, a risk that is highly likely and has a high level of impact would be identified as an area of immediate concern, whereas a risk of lower likelihood or a lower level of exposure would represent a lower level of priority for risk response.

2.7.3 Semiquantitative Risk Assessment

Semiquantitative risk assessment combines the value of qualitative and quantitative risk assessment. A hybrid approach has the realistic input of a qualitative assessment combined with the numerical scale used to determine the impact of a quantitative risk assessment. The goal is to provide a scale with a wide enough range that risk can be assessed in a fairly precise manner, such as 0 to 100. This numeric precision makes it easier to prioritize a large number of risks than is true with a typical qualitative range of three to five levels. However, the risk practitioner should take care to ensure that risk owners do not mistake the origins of these values as coming from purely objective sources.

An example of semiquantitative risk assessment is shown in **figure 2.9**.

Figure 2.9—Assessment Scale—Level of Risk

Qualitative Values	Semi-quantitative Values		Description
Very High	96-100	10	Very high risk means that a threat event could be expected to have multiple severe or catastrophic adverse effects on organizational operations, organizational assets, individuals, other organizations, or the Nation.
High	80-95	8	High risk means that a threat event could be expected to have a severe or catastrophic adverse effect on organizational operation, organizational assets, individuals,other organizations, or the Nation.
Moderate	21-79	5	Moderate risk means that a threat event could be expected to have a serious adverse effect on organizational operations, organizational assets, individuals, or the Nation.
Low	5-20	2	Low risk means that a threat event could be expected to have a limited adverse effect on organizational operations, organizational assets, individuals, other organizations, or the Nation.
Very Low	0-4	0	Very low risk means that a threat event could be expected to have a negligible adverse effect on organizational operations, organizational assets, indivduals, other organizations, or the Nation.

Source: National Institute of Standards and Technology (NIST), *NIST Special Publication 800-30 Revision 1: Guide for Conducting Risk Assessments*, USA, 2012. Reprinted courtesy of the National Institute of Standards and Technology, U.S. Department of Commerce. Not copyrightable in the United States.

Semiquantitative risk assessment can be an effective compromise when impact is quantifiable but likelihood is not. Under such circumstances, applying a basic range of high, medium and low may not offer sufficient precision to generate useful risk ratings, whereas a more granular range of likelihood values may be used with the quantified impact to support specific recommendations for risk response.

2.8 RISK RANKING

The risk practitioner uses the results of risk assessment to place risk in an order that can be used to direct the risk response effort. Risk ranking is derived from a combination of all the components of risk including the recognition of the threats and the characteristics and capabilities of a threat source, the severity of a vulnerability, the likelihood of attack success when considering effectiveness of controls, and the impact to the organization of a successful attack. Taken together, these indicate the level of risk associated with a threat.

Figure 2.10. shows an example of a matrix used to document risk rankings.

Figure 2.10—Template—Adversarial Risk

1	2	3	4	5	6	7	8	9	19	11	12	13
Threat Event	Threat Sources	Threat Source Characteristics			Relevance	Likelihood of Attack Initiation	Vulnerabilities and Predisposing Conditions	Severity and Pervasiveness	Likelihood Initiated Attack Succeeds	Overall Likelihood	Level of Impact	Risk
		Capability	Intent	Targeting								

Source: NIST, *NIST Special Publication 800-30 Revision 1: Guide for Conducting Risk Assessments*, USA, 2012. Reprinted courtesy of the National Institute of Standards and Technology, U.S. Department of Commerce. Not copyrightable in the United States.

Operationally Critical Threat Asset and Vulnerability Evaluation® (OCTAVE®)

OCTAVE is an approach to risk assessment and ranking that is used to assist an organization in understanding, assessing and addressing its information security risk from the perspective of the organization. The OCTAVE methodology is process driven and used to identify, prioritize and manage information security risk. It is intended to help organizations:[3]

- Develop qualitative risk evaluation criteria based on operational risk tolerances
- Identify assets that are critical to the mission of the organization
- Identify vulnerabilities and threats to the critical assets
- Determine and evaluate potential consequences to the organization if threats are realized
- Initiate corrective actions to mitigate risk and create practice-based protection strategy

OCTAVE focuses on critical assets and the risk to those assets using a comprehensive, systematic, context-driven and self-directed evaluation approach. It can help an organization to maintain a proactive security posture and apply the organizational point of view to information security risk management activities.[4]

Some characteristics of OCTAVE are that it:[5]

- Identifies critical information assets
- Focuses risk analysis activities on these critical assets
- Considers the relationships among critical assets, the threats to these assets and the vulnerabilities (both organizational and technological) that can expose assets to threats
- Evaluates risk in operational context (i.e., how the critical assets are used to conduct the organization's business and how they are at risk due to security threats and vulnerabilities)
- Creates practice-based protection strategy for organizational improvement as well as risk mitigation plans to reduce the risk to the organization's critical assets

The OCTAVE process is based on three phases:[6]

- **Phase 1: Build asset-based threat profiles (organizational evaluation)**—The analysis team determines critical assets and what is currently being done to protect them. The security requirements for each critical asset are then identified. Finally, the organizational vulnerabilities with the existing practices and the threat profile for each critical asset are established.
- **Phase 2: Identify infrastructure vulnerabilities (technological evaluation)**—The analysis team identifies network access paths and the classes of IT components related to each critical asset. The team then determines the extent to which each class of component is resistant to network attacks and establishes the technological vulnerabilities that expose the critical assets.

- **Phase 3: Develop security strategy and mitigation plans (strategy and plan development)**—The analysis team establishes risk to the organization's critical assets based on analysis of the information gathered and decides what to do about the risk. The team creates a protection strategy for the organization and mitigation plans to address identified risk. The team also determines the "next steps" required for implementation and gains senior management's approval on the outcome of the entire process.

2.8.1 Risk Appetite Bands

After risk has been determined using quantitative, qualitative or semiquantitative methods, the risk practitioner needs to review whether the risk levels are within the boundaries of acceptable risk as defined by senior management through the setting of the organizational risk appetite and risk tolerance.

Figure 2.11 illustrates levels of risk on the basis of magnitude and frequency (impact and likelihood, respectively):
- Risk that falls within the risk appetite is "acceptable."
- Risk that is outside of the risk appetite but within the risk tolerance is "unacceptable."
- Risk that is outside of the risk tolerance is "really unacceptable."

Figure 2.11—Risk Map Indicating Risk Appetite Bands

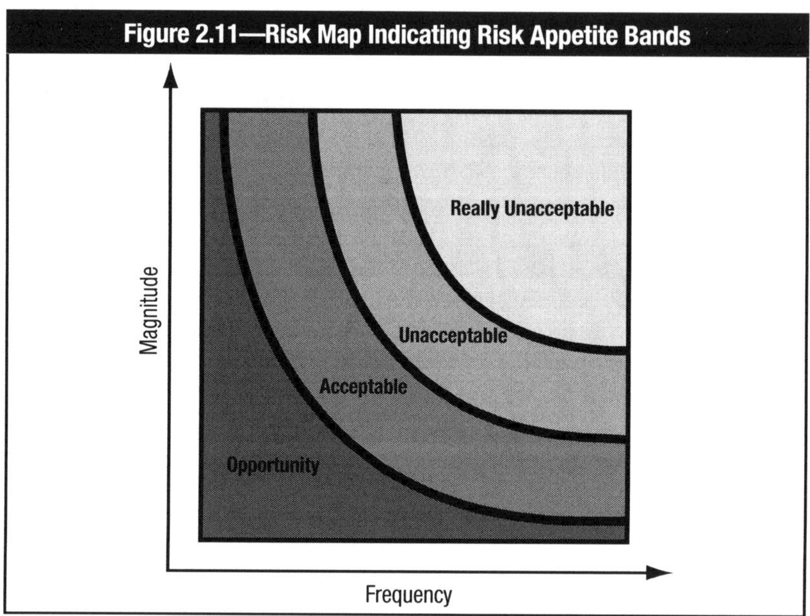

2.8.2 Risk Ownership and Accountability

Each risk must be linked to an individual who accepts ownership of the risk. The risk owner is tasked with making the decision of what the best response is to the identified risk and must be at a level in the organization where he or she is authorized to make decisions on behalf of the organization and can be held accountable for those decisions. To ensure accountability, the ownership of risk must be with an individual, not with a department or the organization as a whole.

2.9 DOCUMENTING RISK ASSESSMENTS

At the conclusion of the IT risk assessment phase, the risk practitioner compiles the results of the risk assessment into a comprehensive report for senior management. The risk assessment report should indicate any gaps between the current risk environment and the desired state of IT risk, advise whether these gaps are within acceptable levels, and provide some basis on which to judge the severity of the identified issue. The risk assessment report should document the process used as well as the result of the risk assessment. Each risk should be documented in a manner that is understandable to management and clearly states the risk levels and priorities.

The risk assessment report may be adjusted according to the needs of the organization and the direction of senior management. A typical risk assessment report includes the following sections:
• Objectives of the risk assessment process
• Scope and description of the area subject to assessment
• External context and factors affecting risk
• Internal factors or limitations affecting risk assessment
• Risk assessment criteria
• Risk assessment methodology used
• Resources and references used
• Identification of risk, threats and vulnerabilities
• Assumptions used in the risk assessment
• Potential of unknown factors affecting assessment
• Results of risk assessment
• Recommendations and conclusions

The risk assessment should be performed in a consistent manner that supports future risk assessment efforts and would provide predictable results. All risk should be noted in the report, including issues that may already have been addressed, in order to provide an accurate picture of risk to senior management. This comprehensive approach ensures that if a control that was in place at the time of the assessment is later removed, the risk practitioner is not suspected of overlooking it during the identification and assessment processes.

Organizations assess IT risk because IT risk is a form of business risk, not because it is a nuisance to the IT staff. The risk practitioner should refrain from using terminology that is specific to IT or may be misinterpreted by management. Many terms and acronyms commonly used in IT have different meaning to other disciplines in the organization, which may result in the misunderstanding of the problem and the proposed solutions. The risk practitioner should ensure that the report is clear, concise and accurate and is free from terminology that could be misunderstood or is subject to misinterpretation.

2.9.1 Addressing Bypassed Risk

The risk practitioner should take care to ensure that all IT risk is either evaluated or intentionally bypassed. Some IT risk events apply only to organizations that meet particular criteria. For instance, not all regions are subject to the same natural disasters. Likewise, the risk of losing credit card data applies only to those organizations that store or process credit card data. When an assumption is made to not include a particular risk in an assessment, the risk practitioner may find it beneficial to document that the risk was intentionally bypassed rather than missed in the risk assessment process. In general, this occurs during the risk identification process, but as a deeper level of investigation and evaluation of risk is conducted, new types of risk may surface while others that were previously overlooked are found to be relevant. For this reason, the risk practitioner should reevaluate each documented risk to ensure that it was identified and assessed accurately based on the current risk landscape.

2.9.2 Updating the Risk Register

Risk management is an ongoing process, and the risk register is a living document that must be continuously updated. The risk practitioner should update the risk register with the information gathered in the assessment process as the last step before proceeding to the next phase of risk management.

2.10 SUMMARY

The risk practitioner has a responsibility to assess each risk facing the organization in terms of both likelihood and impact and rank the results for appropriate response. Much of this work is based on the results of the risk identification phase, but the risk practitioner should also validate the work of the previous phase and ensure that, as much as possible, all risk is identified, assessed, documented and reported to senior management.

ENDNOTES

[1] ISO; *IEC 31010:2009 Risk Management—Risk Assessment Techniques*, Switzerland, 2009

[2] Kissel, Richard; Kevin Stine; Matthew Scholl; Hart Rossman; Jim Fahlsing; Jessica Gulick; *NIST Special Publication 800-64 Revision 2: System Considerations in the System Development Life Cycle*, National Institute of Standards and Technology, USA, 2008

[3] Panda, Parthajit, "The OCTAVE Approach to Information Security Risk Assessment," *ISACA Journal*, Volume 4, 2009

[4] *Ibid.*

[5] *Ibid.*

[6] *Ibid.*

Chapter 3: Risk Response and Mitigation

Section Two: Contents

Section One: Overview

DOMAIN DEFINITION

Determine risk response options and evaluate their efficiency and effectiveness to manage risk in alignment with business objectives.

LEARNING OBJECTIVES

The objective of this domain is to ensure that the CRISC candidate has the knowledge necessary to:
- List the different risk response options
- Define various parameters for risk response selection
- Explain how residual risk relates to inherent risk, risk appetite and risk tolerance
- Discuss the need for performing a cost-benefit analysis when determining a risk response
- Develop a risk action plan
- Explain the principles of risk ownership
- Leverage an understanding of the system development life cycle (SDLC) process to implement IS controls efficiently and effectively
- Understand the need for control maintenance

CRISC EXAM REFERENCE

This domain represents 23 percent of the CRISC exam (approximately 35 questions).

TASK AND KNOWLEDGE STATEMENTS

TASKS

There are seven tasks within this domain that a CRISC candidate must know how to perform. These relate to IT risk identification.

T3.1 Consult with risk owners to select and align recommended risk responses with business objectives and enable informed risk decisions.
T3.2 Consult with, or assist, risk owners on the development of risk action plans to ensure that plans include key elements (e.g., response, cost, target date).
T3.3 Consult on the design and implementation or adjustment of mitigating controls to ensure that the risk is managed to an acceptable level.
T3.4 Ensure that control ownership is assigned in order to establish clear lines of accountability.
T3.5 Assist control owners in developing control procedures and documentation to enable efficient and effective control execution.
T3.6 Update the risk register to reflect changes in risk and management's risk response.
T3.7 Validate that risk responses have been executed according to the risk action plans.

KNOWLEDGE STATEMENTS

The CRISC candidate should be familiar with the task statements relevant to each domain in the CRISC job practice. The tasks are supported by 41 knowledge statements that delineate each of the areas in which the risk practitioner must have a good understanding in order to perform the tasks. Many knowledge statements support tasks that cross domains.

The CRISC candidate should have knowledge of:
1. Laws, regulations, standards and compliance requirements
2. Industry trends and emerging technologies
3. Enterprise systems architecture (e.g., platforms, networks, applications, databases and operating systems)
4. Business goals and objectives
5. Contractual requirements with customers and third-party service providers

6. Threats and vulnerabilities related to:
   6.1 Business processes and initiatives
   6.2 Third-party management
   6.3 Data management
   6.4 Hardware, software and appliances
   6.5 The system development life cycle (SDLC)
   6.6 Project and program management
   6.7 Business continuity and disaster recovery management (DRM)
   6.8 Management of IT operations
   6.9 Emerging technologies
7. Methods to identify risk
8. Risk scenario development tools and techniques
9. Risk identification and classification standards, and frameworks
10. Risk events/incident concepts (e.g., contributing conditions, lessons learned, loss result)
11. Elements of a risk register
12. Risk appetite and tolerance
13. Risk analysis methodologies (quantitative and qualitative)
14. Organizational structures
15. Organizational culture, ethics and behavior
16. Organizational assets (e.g., people, technology, data, trademarks, intellectual property) and business processes, including enterprise risk management (ERM)
17. Organizational policies and standards
18. Business process review tools and techniques
19. Analysis techniques (e.g., root cause, gap, cost-benefit, return on investment [ROI])
20. Capability assessment models and improvement techniques and strategies
21. Data analysis, validation and aggregation techniques (e.g., trend analysis, modeling)
22. Data collection and extraction tools and techniques
23. Principles of risk and control ownership
24. Characteristics of inherent and residual risk
25. Exception management practices
26. Risk assessment standards, frameworks and techniques
27. Risk response options (i.e., accept, mitigate, avoid, transfer) and criteria for selection
28. Information security concepts and principles, including confidentiality, integrity and availability of information
29. Systems control design and implementation, including testing methodologies and practices
30. The impact of emerging technologies on design and implementation of controls
31. Requirements, principles, and practices for educating and training on risk and control activities
32. Key risk indicators (KRIs)
33. Risk monitoring standards and frameworks
34. Risk monitoring tools and techniques
35. Risk reporting tools and techniques
36. IT risk management best practices
37. Key performance indicator (KPIs)
38. Control types, standards, and frameworks
39. Control monitoring and reporting tools and techniques
40. Control assessment types (e.g., self-assessments, audits, vulnerability assessments, penetration tests, third-party assurance)
41. Control activities, objectives, practices and metrics related to:
   41.1 Business processes
   41.2 Information security, including technology certification and accreditation practices
   41.3 Third-party management, including service delivery
   41.4 Data management
   41.5 The system development life cycle (SDLC)
   41.6 Project and program management
   41.7 Business continuity and disaster recovery management (DRM)
   41.8 IT operations management
   41.9 The information systems architecture (e.g., platforms, networks, applications, databases and operating systems)

SELF-ASSESSMENT QUESTIONS

CRISC self-assessment questions support the content in this manual and provide an understanding of the type and structure of questions that have typically appeared on the exam. Questions are written in a multiple-choice format and designed for one best answer. Each question has a stem (question) and four options (answer choices). The stem may be written in the form of a question or an incomplete statement. In some instances, a scenario or a description problem may also be included. These questions normally include a description of a situation and require the candidate to answer two or more questions based on the information provided. Many times a question will require the candidate to choose the **MOST** likely or **BEST** answer among the options provided.

In each case, the candidate must read the question carefully, eliminate known incorrect answers and then make the best choice possible. Knowing the format in which questions are asked, and how to study and gain knowledge of what will be tested, will help the candidate correctly answer the questions.

3-1 When a risk cannot be sufficiently mitigated through manual or automatic controls, which of the following options will **BEST** protect the enterprise from the potential financial impact of the risk?

A. Insuring against the risk
B. Updating the IT risk register
C. Improving staff training in the risk area
D. Outsourcing the related business process to a third party

3-2 To be effective, risk mitigation **MUST** reduce the:

A. residual risk.
B. inherent risk.
C. frequency of a threat.
D. impact of a threat.

3-3 The **BEST** control to prevent unauthorized access to an enterprise's information is user:

A. accountability.
B. authentication.
C. identification.
D. access rules.

3-4 Which of the following controls **BEST** protects an enterprise from unauthorized individuals gaining access to sensitive information?

A. Using a challenge response system
B. Forcing periodic password changes
C. Monitoring and recording unsuccessful logon attempts
D. Providing access on a need-to-know basis

3-5 Which of the following defenses is **BEST** to use against phishing attacks?

A. An intrusion detection system (IDS)
B. Spam filters
C. End-user awareness
D. Application hardening

3-6 When responding to an identified risk event, the **MOST** important stakeholders involved in reviewing risk response options to an IT risk are the:

A. information security managers.
B. internal auditors.
C. incident response team members.
D. business managers.

3-7 Which of the following choices should be considered **FIRST** when designing information system controls?

A. The organizational strategic plan
B. The existing IT environment
C. The present IT budget
D. The IT strategic plan

3-8 Residual risk can be accurately calculated on the basis of:

A. Threats and vulnerabilities
B. Inherent risk and control risk
C. Compliance risk and reputation
D. Risk governance and risk response

ANSWERS TO SELF-ASSESSMENT QUESTIONS

Correct answers are shown in **bold**.

3-1 **A. An insurance policy can compensate the enterprise monetarily for the impact of the risk by transferring the risk to the insurance company.**
B. Updating the risk register (with lower values for impact and probability) will not actually change the risk, only management's perception of it.
C. Staff capacity to detect or mitigate the risk may potentially reduce the financial impact, but insurance allows for the risk to be completely mitigated.
D. Outsourcing the process containing the risk does not necessarily remove or change the risk.

3-2 **A. The objective of risk reduction is to reduce the residual risk to levels below the enterprise's risk tolerance level.**
B. Reduction of inherent risk is not the goal of typical risk reduction/mitigation efforts.
C. Risk reduction efforts can focus on reducing either frequency or impact.
D. Risk reduction efforts can focus on reducing either frequency or impact.

3-3 A. User accountability does not prevent unauthorized access; it maps a given activity or event back to the responsible party.
B. Authentication verifies the user's identity and the right to access information according to the access rules.
C. User identification without authentication does not grant access.
D. Access rules without identification and authentication do not grant access.

3-4 A. Verifying the user's identification through a challenge response does not completely address the issue of access risk if access was not appropriately designed in the first place.
B. Forcing users to change their passwords does not guarantee that access control is appropriately assigned.
C. Monitoring unsuccessful access logon attempts does not address the risk of appropriate access rights.
D. Physical or logical system access should be assigned on a need-to-know basis (legitimate business requirements) and in ways that incorporate least privilege and segregation of duties (SoD).

3-5 A. An intrusion detection system (IDS) does not protect against phishing attacks because phishing attacks usually do not have the same patterns or unique signatures.
B. While certain highly specialized spam filters can reduce the number of phishing emails that reach their addressees' inboxes, they are not as effective in addressing phishing attacks as end-user awareness.
C. Phishing attacks are a type of social engineering attack and are best defended by end-user awareness training.
D. Application hardening does not protect against phishing attacks because phishing attacks generally use email as the attack vector, with the end user, not the application, as the vulnerable point.

3-6 A. Business managers are accountable for managing the associated risk and will determine what actions to take based on the information provided by others, which may include collaboration with and support from IT security managers.
B. Risk response is not a function of internal audit.
C. The incident response team must ensure open communication to management and stakeholders to ensure that business managers/leaders understand the associated risk and are provided enough information to make informed risk-based decisions.
D. Business managers are accountable for managing the associated risk and will determine what actions to take based on the information provided by others.

3-7 **A. Review of the enterprise's strategic plan is the first step in designing effective IS controls that would fit the enterprise's long-term plans.**
B. Review of the existing IT environment, although useful and necessary, is not the first task that needs to be undertaken.
C. The present IT budget is one of the components of the strategic plan.
D. The IT strategic plan exists to support the enterprise's strategic plan.

3-8 A. Although threat and vulnerability are elements of inherent risk, which is one factor needed to calculate residual risk, the risk practitioner does not necessarily need to have the components of inherent risk in order to know the inherent risk.
B. Inherent risk multiplied by control risk is the formula to calculate residual risk.
C. Compliance risk is the current and prospective risk to earnings or capital arising from violations of, or nonconformance with, laws, rules, regulations, prescribed practices, internal policies and procedures, or ethical standards. Compliance risk can lead to reputational damage, but having these factors does not provide a basis for calculating residual risk.
D. Risk governance and risk response are risk domains, not risk elements.

Note: For more self-assessment questions, you may also want to obtain a copy of the *CRISC™ Review Questions, Answers & Explanations Manual 4th Edition* or the Database, which each consist of 500 multiple-choice study questions, answers and explanations.

SUGGESTED RESOURCES FOR FURTHER STUDY

In addition to the resources cited throughout this manual, the following resources are suggested for further study in this domain (publications in **bold** are stocked in the ISACA Bookstore):

Abkowitz, Mark D.; *Operational Risk Management: A Case Study Approach to Effective Planning and Response*, John Wiley & Sons, USA, 2008

International Organization for Standardization (ISO)/International Electrotechnical Commission (IEC); *ISO/IEC 27001:2013 Information technology—Security techniques—Information security management systems—Requirements*, Switzerland, 2013

ISO/IEC; *ISO/IEC 27005:2011— Information Security Risk Management*, Switzerland, 2011

ISO/IEC; *ISO/IEC 31000:2009 Risk management —Principles and Guidelines*, Switzerland, 2009

ISACA, *COBIT® 5 for Risk*, USA, 2013, *www.isaca.org/cobit*

ISACA, *The Risk IT Framework*, USA, 2009

ISACA, *The Risk IT Practitioner Guide*, USA, 2009

The Open Web Application Security Project, www.owasp.org

Section Two: Content

3.0 OVERVIEW

The risk response phase of risk management, shown in **figure 3.1**, focuses on the decisions made regarding the correct way to respond to risk. Risk response is based on the information provided in the earlier steps of risk identification and risk assessment, but that information is balanced with the constraints placed on the organization through budget, time, resources, strategic plans, regulations, customer expectations and other business factors. Management must be prepared to justify its risk response decision to stakeholders and provide a road map to implementing agreed-upon changes according to a reasonable schedule. Risk response must also be undertaken in a way that protects operations without unduly impairing them, making the careful selection of controls that are put in place to address risk an important part of the process.

Figure 3.1—The IT Risk Management Life Cycle

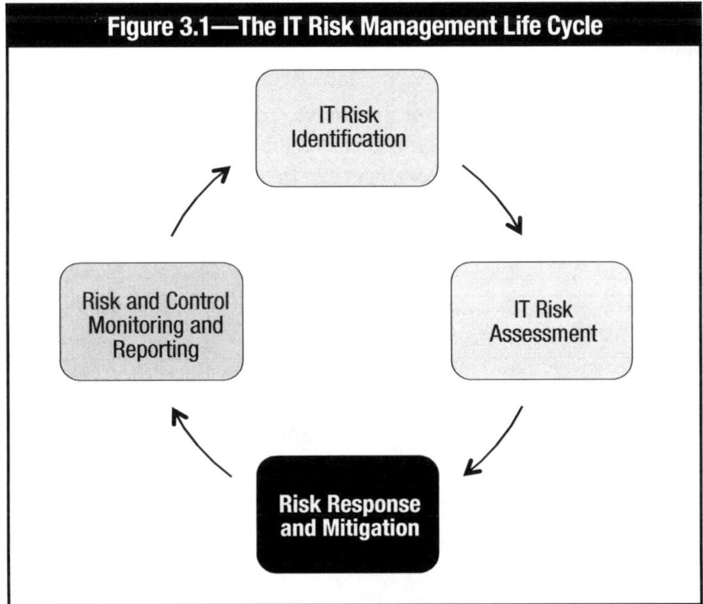

3.1 ALIGNING RISK RESPONSE WITH BUSINESS OBJECTIVES

The risk assessment report and risk register document the risk identified during the identification and assessment phases of the risk management process. Both the report and the register should indicate the assessed level or priority of each risk. The recommendations in the report are guidelines, and several recommendations may be provided. Management is responsible for evaluating and responding to the recommendations included in the report, first by determining the best response and then developing an action plan and implementation strategy that addresses the risk in a manner consistent with the risk appetite and risk tolerance of the organization.

The organization exists to meet its mission. Management must always be aware of the drivers for risk management, such as compliance with regulations and the need to support and align the risk response with business priorities and objectives. A risk response that would negatively impact the ability of the organization to meet its mission must be considered very carefully.

3.2 RISK RESPONSE OPTIONS

Evaluation of an appropriate risk response is part of the risk management process cycle, not a one-time effort. There are four commonly accepted options for risk response:

- Risk acceptance
- Risk mitigation
- Risk sharing (transfer)
- Risk avoidance

The purpose of defining a risk response is to bring risk in line with the defined risk tolerance of the organization as cost-effectively as possible, not to eliminate or minimize the risk at all costs.

3.2.1 Risk Acceptance

The choice to accept risk is a conscious decision made by senior management to recognize the existence of risk and knowingly decide to allow (assume) the risk to remain without (further) mitigation. Management is responsible for the impact of a risk event should it occur, so the decision to accept a risk is made according to the risk appetite and risk tolerance set by senior management.

Because the goal of risk management is to bring risk within acceptable levels as cost-effectively as possible, risk that falls within the organizational risk appetite should be accepted. Risk tolerance is used as an exceptional circumstance where the level of risk may exceed the risk acceptance boundary set by senior management, but the decision is made to accept the risk anyway. Risk tolerance is typically a factor when either no controls are available or the cost of the controls would outweigh their benefits. Self-insurance, or deciding to absorb the potential costs of an incident, is another form of risk acceptance.

Examples of risk acceptance as a result of risk tolerance include:

- Projections indicate that a certain project will not deliver the required business functionality by the planned delivery date, but management decides to proceed with the project.
- A particular risk would have catastrophic consequences but is assessed to be extremely rare, and approaches to reduce it are cost-prohibitive. Management may decide to accept this risk.

Acceptance of risk must be done by someone empowered to act on behalf of the organization at a level commensurate with the projected consequences associated with the risk. Risk acceptance is not the same as risk ignorance, which refers to both the failure to identify or acknowledge risk and the blind acceptance of risk without knowing or acknowledging what the risk level really is. Too often, acceptance of risk is based on poorly calculated risk levels, and many organizations have found, in the wake of an incident, that the actual level of risk accepted was far greater than what they had intended to accept. Careful review of actuarial data or the outcomes of similar incidents at other organizations and their resulting impact may help the risk practitioner to estimate true likely incident costs.

In addition to acceptance of risk at the outset (which relates to inherent risk), organizations should also recognize that other approaches to risk management typically yield residual risk that must itself be acceptable to the organization. A mitigated risk, for instance, is still non-zero and must be evaluated to ensure that it falls within the organizational risk appetite.[1] Additionally, risk acceptance is not a one-time process. The level of risk and potential impact of an incident can change dramatically as assets increase or decrease in value, new threats emerge, attackers gain additional skills or motivation, vulnerabilities are discovered, the existing controls become weaker or the operations team becomes less diligent. Organizations should conduct regular reviews of their accepted risk to ensure that these continue to be consistent with senior management's level of risk tolerance.

3.2.2 Risk Mitigation

Risk mitigation refers to actions that the organization takes in order to reduce a risk. Mitigation is typically achieved through security controls, which affect the frequency and/or impact of the risk.

Some examples of risk mitigation are:

- Strengthening overall risk management practices, such as implementing sufficiently mature risk management processes
- Deploying new technical, management or operational controls that reduce either the likelihood or the impact of an adverse event
- Installing a new access control system
- Implementing policies or operational procedures
- Developing an effective incident response and business continuity plan (BCP)
- Using compensating controls

Realistic choices for risk mitigation may be determined in part by the scope of the risk or the speed at which an incident can expand. In some cases, multiple controls may be needed in order to reduce risk to an acceptable level. Acceptance of residual risk should be documented and subject to periodic reevaluation in the same manner as applies to the acceptance of inherent risk.

3.2.3 Risk Transfer (Sharing)

Risk transfer is a decision to reduce loss by having another organization incur the cost. The most common example of risk transfer is the purchasing of insurance, which provides a guarantee of compensation or replacement should a loss occur. Partnerships are another form of risk transfer, in which two or more organizations work together under an arrangement in which both risk of loss and potential for profit are divided among the participants according to agreed-upon terms and conditions.

The term "risk transfer" can be misleading, because an organization that sustains a substantial loss may accrue costs beyond what can be reimbursed by insurance, particularly in the form of damage to its brand (reputational risk). For instance, a major financial institution that exposes large amounts of confidential customer data might be reimbursed for immediate expenses by an insurance policy, but customers are unlikely to be impressed by the insurance payout and may take their business elsewhere, and these long-term costs are typically not covered by insurance. Risk transfer is a legitimate option for responding to risk, but risk practitioners and managers should take care to remember that stakeholders rarely regard the transfer of risk as complete absolution of blame.

Transferred risk should be reviewed on a regular basis to ensure that it remains appropriate and adequate. For example, an organization should ensure that the current amount of insurance is adequate to cover losses and that the organization is compliant with the terms and conditions of the coverage. Risk transfer arrangements are only effective when they are properly maintained, and the moment after an incident occurs is the worst time to realize that such arrangements recently expired.

3.2.4 Risk Avoidance

Risk avoidance means exiting the activities or conditions that give rise to risk. Risk avoidance is the choice that remains when no other response is adequate, meaning all of the following are true:

- The exposure level is deemed unacceptable by management.
- The risk cannot be transferred.
- Mitigation that would bring the risk in line with acceptable levels is either impossible or would cost more than the benefits that the organization derives from the activities.

Some examples of risk avoidance include:

- Relocating a data center away from a region with significant natural hazards
- Evacuating staff from a work area in which there is substantial civil unrest
- Declining a project whose business case shows a substantial likelihood of failure
- Rejecting a partnership agreement in which potential losses are allocated to your organization, but the partner stands to benefit from most potential profits
- Deciding not to use proprietary technology or software from a vendor because it has only recently begun operations or is undergoing bankruptcy

Risk avoidance is not a decision to make lightly, but it may be the best choice in some instances. The role of the risk practitioner is to provide management with timely, accurate risk evaluations and solid supporting data so that informed decisions can be made. In addition, the risk practitioner should be ready to refresh the evaluations and data whenever management chooses to review earlier decisions for potential changes in strategy or strategic plans.

3.3 ANALYSIS TECHNIQUES

The determination of the best risk response can be based on one of several analysis techniques. The risk response chosen by the management of the organization is often based on the evaluation of several possible response options. The management of the organization should consider several factors in selecting a response, such as:

- The priority of the risk as indicated in the risk assessment report
- The recommended controls from the risk assessment report

- Any other response alternatives that are suggested through further analysis
- The cost of the various response options, including:
  - Acquisition cost
  - Training cost
  - Impact of productivity
  - Maintenance and licensing costs
- Requirements for compliance with regulations or legislation
- Alignment of the response option with the strategy of the organization
- Possibility of integrating the response with other organizational initiatives
- Compatibility with other controls in place
- Time, resources and budget available

The selection of the appropriate response is based primarily on the calculation of value obtained for cost incurred. In other words, does the cost of implementing a specific risk response provide enough value to the organization for it to be a wise decision? In the end, the decision may be to accept the risk or to respond to the risk through a combination of managerial, technical and physical controls.

Specific responses to risk are typically recommended in business cases, which document the problems and one or more options as well as providing detailed analyses on which management may base decisions. Beyond being common and useful tools for decision support, business cases also serve as operational tools against which to evaluate and support investments throughout their life cycles. Two of the most common forms of analysis used to prepare a business case for risk response are cost-benefit analysis and return on investment (ROI).

3.3.1 Cost-benefit Analysis

Cost-benefit analysis is used to justify the expense associated with the implementation of controls. The expenditure on a control cannot be justified if the benefit realized from the control is less than the cost. There are several factors that must be included in calculating the total cost of the control:

- Cost of acquisition
  - Evaluation of solutions
  - Cost of the control
  - Cost of training
  - Cost to rearchitect systems
- Ongoing cost of maintenance
  - License costs
  - Cost of staff to monitor and report on control
  - Impact on productivity/performance
  - Cost of support and technical assistance
- Cost to remove/replace control

Likewise, the benefit realized from the control must consider several factors:

- Reduced cost of risk event
- Reduced liability
- Reduced insurance premiums
- Increased customer confidence
- Increased shareholder confidence
- Trust from financial backers
- Faster recovery
- Better employee relations (safety)

The costs and benefits may be calculated using both qualitative and quantitative measures. The impact is often related to the length of the outage, the frequency of the outage (i.e., customers may more readily forgive one incident than frequent problems), and scope and publicity of the incident, but the underlying value is the extent to which a given expense results in reduction of risk.

3.3.2 Return on Investment

Many organizations justify an investment in projects, tools or new ventures using ROI, which calculates how long it takes to recoup its cost through value added or other savings produced. A new computer system, for example, might pay for itself over three years as a result of better productivity, lower numbers of staff required or increased sales.

Calculating the ROI associated with the implementation of a control is often difficult, in part because it depends on predicting the likelihood of a successful attack. An additional complication is the goal of a control is to bring risk to an acceptable level rather than eliminating it outright. Some organizations use the term return on security investment (ROSI) to refer to ROI specifically in relation to the payback for security controls. In determining ROI or ROSI, the organization is trying to forecast the likelihood and impact of an incident and deciding what is an adequate level of protection. The risk practitioner should bear in mind that the amount of security an organization decides to implement is dependent on its appetite for risk and the perception of exposure.

3.4 VULNERABILITIES ASSOCIATED WITH NEW CONTROLS

Implementing a new or modified control that provides benefit to the organization may also introduce new vulnerabilities. For example, an access control system that protects against unauthorized access may inadvertently deny access to legitimate users who forget passwords or lose access cards, resulting in calls for technical assistance, delays in processing and user frustration. In some cases, the risk associated with deploying a particular control may exceed the risk that it is meant to address. Organizations should coordinate with stakeholders on a proactive basis and perform rigorous user acceptance testing (UAT) under conditions as close to real-world use as possible prior to full implementation.

3.5 DEVELOPING A RISK ACTION PLAN

There are several responses to risk that may be considered, and the risk practitioner plays a consultative role in assisting risk owners with deciding on the correct response to a risk (i.e., to accept, transfer mitigate or avoid a risk). The ultimate decision on risk response lies with the risk owner, but the risk practitioner can and should provide advice on technologies, policies, procedures, control effectiveness and leveraging of existing controls.

The decision plan to implement a control is based on many factors, including:
• Current risk level
• Applicable laws and regulations
• Ongoing projects
• Strategic plans and management priorities
• Current and projected budgets
• Availability of staff
• Public pressure
• Actions of competitors

The risk response process shown in **figure 3.2** reflects the following concepts:
1. Risk scenarios (from risk identification) drive the risk analysis and assessment.
2. Risk analysis (from risk assessment) leads to the documentation (mapping) of risk.
3. Risk response is determined by the risk appetite of the organization. If the risk exceeds the appetite, then a risk response to address the risk must be considered.
4. The risk response options are considered along with the risk response parameters to determine the best available risk response.
5. The selected risk response is documented.
6. The risk responses are prioritized according to the current risk environment and the cost-benefit of risk mitigation.
7. A risk action plan is created to manage the risk response projects.

Figure 3.2—Control Category Interdependencies

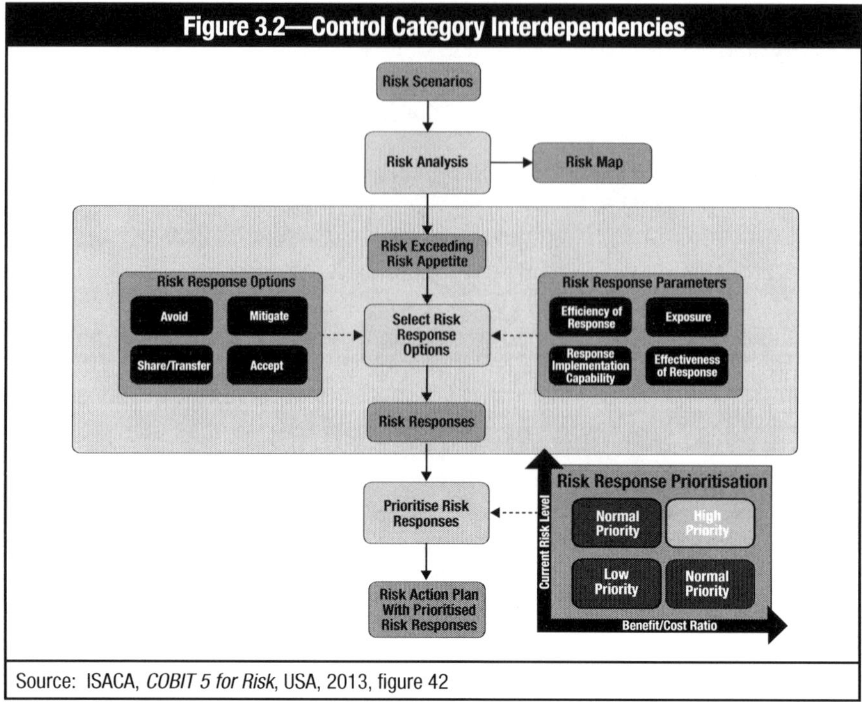

Source: ISACA, *COBIT 5 for Risk*, USA, 2013, figure 42

Risk response is subject to the strategic direction and technological direction of the organization, which can affect both the choice of one control over another and its schedule for deployment.

These decisions can be difficult to make and often require some method of comparing control options. Questions that the risk practitioner should keep in mind include:

- Would the control be effective?
- Would the control provide a satisfactory ROI (in organizations where this is a driving concern)?
- Does the organization have sufficient skill to implement, configure and maintain the control?
- Is there sufficient budget and time to implement the control?
- How much would the control cost to operate on an annual basis, or what would the impact of the control be on productivity?

Risk action planning should be run as a project, with a defined start and end date. The end date is often used to determine the critical path of the project, which refers to those elements of the project that may have a direct impact on whether the end date can be met. The risk practitioner may not be a project manager, but he/she should understand that change in the delivery of any project element on the critical path affects the delivery of the entire project; for example, a project that does not receive its equipment from the supplier on time may not be able to meet the scheduled project dates. Critical path elements should be given special consideration for timeliness, because delays in these elements increase overall project risk. Through experience and careful evaluation, the risk practitioner can advise the risk owner on the feasibility of project dates, the expected workload associated with the project, the costs of the project and the overall success of the project according to risk management and business goals.

3.6 BUSINESS PROCESS REVIEW TOOLS AND TECHNIQUES

A business process review examines the effectiveness and efficiency of an organization in meeting its goals and objectives. When responding to risk, the organization should review its business processes to ensure that the correct solution is selected to integrate and work effectively with the existing environment, or whether the environment should be reengineered to improve its efficiency and success in meeting organizational objectives.

A business process review requires input from knowledgeable representatives from all affected departments within the organization, and it may also bring in external experts who can provide advice and assistance. The purpose of the business process review is to identify ways to:

- Identify problems or issues with the current process

- Gather information toward improving processes
- Prepare a road map to implement required changes
- Assign responsibility and accountability for projects
- Schedule individual projects according to priority
- Monitor project progress for attainment of milestones and production of deliverables
- Review and obtain feedback on project results

The steps of a business process review are:

1. Document and evaluate current business processes
   a. List critical processes, supply chains and services
      i. Classify critical processes by granularity level (e.g., first, second or third degree)
      ii. Identify responsibility and accountability for each process
   b. Document current business processes and risk
      i. Review documentation
      ii. Interview management, users and other stakeholders
      iii. Observe actual processes
      iv. Validate with business representatives
         1. Validate training and skills required of personnel
   c. Document current issues and problems
   d. Baseline other organizations
   e. Work with team members to discover potential solutions and improvements
2. Identify potential changes
   a. Use focus groups and workshops to determine process improvements
   b. Validate proposed changes with management and obtain approval to proceed
3. Schedule and implement changes
   a. Design changes
   b. Identify dependencies
   c. Communicate the schedule of changes
4. Feedback and evaluation
   a. Measure operational efficiencies
      i. Customer satisfaction
      ii. User satisfaction
      iii. Improvements in productivity/quality
      iv. Feedback to improve processes

3.7 CONTROL DESIGN AND IMPLEMENTATION

Risk mitigation is accomplished through the use of controls. Controls may be proactive, meaning that they attempt to prevent an incident. Controls may also be reactive, meaning that they allow the detection, containment and recovery from an incident. Proactive controls are often called safeguards, and reactive controls are known as countermeasures. For example, a sign that warns a person about a dangerous condition is a safeguard, whereas a fire extinguisher or sprinkler system is a countermeasure.

Every organization has some controls in place, and the risk assessment should document those controls and calculate their effectiveness in mitigating risk. In some cases, the controls may be sufficient, whereas in others, the controls may need adjustment or replacement. An effective control is one that prevents, detects and/or contains an incident and enables recovery from a risk event.

The risk practitioner typically provides advice on the selection, design, implementation, testing and operation of the controls.

It is common for an organization to have some situations where the controls currently in place are not sufficient to adequately protect the organization. In most cases, this requires the adjustment of the current controls or the implementation of new controls. However, it may not be feasible to reduce the risk to an acceptable level by either adjusting or implementing controls due to reasons such as cost, job requirements or availability of controls. An example of this could be found in a small organization when an individual is given administrator rights on a system and there is not adequate segregation of duties (SoD). In this

case, it may not be feasible to implement a new or enhanced control; some personnel need administrator rights to perform their jobs, and the risk cannot justify the cost of hiring new staff to address SoD. Therefore, the risk owner may consider implementing compensating controls to reduce the risk. Compensating controls address the weaknesses in the existing controls through concepts such as layered defense, increased supervision, procedural controls, or increased audits and logging of system activity. These measures will work to compensate for the risk that could not be addressed in other ways.

Controls may be grouped into managerial, technical or physical controls, and within each of those groups of controls are various types of controls that can be used, such as preventive, detective and corrective, recovery and compensating controls. Controls are discussed in more detail in chapter 2 IT Risk Assessment. An example of a control matrix is shown in **figure 3.3**. Many controls may fit into more than one category.

Figure 3.3—Control Matrix

	Managerial	Technical	Physical
Compensating	Separation of duties (SoD)	Two-factor authentication	Dual control operations
Corrective	Remove access	Network isolation	Close fire doors
Detective	Audit	Intrusion detection system (IDS)	Motion sensor
Deterrent	Disciplinary policy	Warning banner on login	"Beware of Dog" sign
Preventive	User registration process	Login screen	Fence

The interaction between the control types is shown in **figure 3.4**.

Figure 3.4:—Control Category Interdependencies

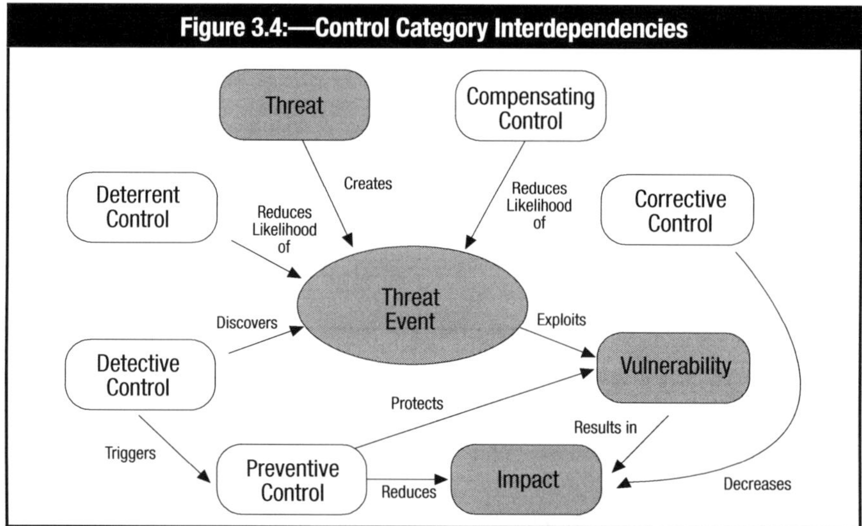

3.7.1 Control Standards and Frameworks

Selection of controls requires evaluation and implementation. Based on data collected through appropriate analysis methods (e.g., cost-benefit analysis, ROI, etc.), management decides on the best available control or group of controls to mitigate a specific risk. However, a poorly implemented control may end up posing a risk to the organization if it creates a false sense of security or causes other controls to function less effectively. In particular, the implementation of technical control requires proper procedures and training, and all controls should be assigned to specific owners accountable for ensuring that they are monitored and tested on a regular basis.

Many industries have standards that may be beneficial as benchmarks for security within their respective sectors even when compliance is not required by law. For instance, the Payment Card Industry Data Security Standard (PCI DSS) is used by all organizations that process debit or credit cards. Similar standards are found in the health care, accounting, audit and telecommunications industries. In some industries, voluntary standards are supplemented by regulations that require compliance. Standards may also include or be supported by frameworks, which provide sets of fundamental controls whose adoption is recommended, but whose implementation is left up to the organization.

3.7.2 Administrative, Technical and Physical Controls

Controls are often categorized as managerial (administrative), technical and physical, as shown in **figure 3.5**.

Figure 3.5—Control Methods

Category	Description
Managerial (administrative)	These are related to the oversight, reporting, procedures and operations of a process. These include controls such as policy, procedures, balancing, employee development and compliance reporting.
Technical	Technical controls are also known as logical controls and are provided through the use of a technology, piece of equipment or device. Examples of technical controls include firewalls, network or host-based intrusion detection systems (IDSs), passwords and antivirus software. A technical control requires proper managerial (administrative) controls to operate correctly.
Physical	Physical controls are locks, fences, closed-circuit TV (CCTV) and devices that are installed to physically restrict access to a facility or hardware. Physical controls require maintenance, monitoring and the ability to assess and react to an alert should a physical control indicate a problem.
Source: ISACA, *CISA Review Manual 26th Edition*, USA, 2015, figure 5.5	

3.8 CONTROL MONITORING AND EFFECTIVENESS

To support the ability to monitor and report on risk, the risk practitioner should ensure that processes, logs and audit hooks are placed into the control framework. This allows for the monitoring and evaluation of controls. As controls are designed, implemented and operated, the risk practitioner should ensure that logs are enabled, controls are able to be tested and regular reporting procedures are developed.

3.8.1 Control Monitoring and Reporting Tools and Techniques

The actual monitoring and reporting of controls should be performed in the risk monitoring phase of risk management; however, the risk practitioner should ensure that the capability to monitor a control and to support monitoring systems is addressed in control design. If the organization is using a managed security service provider (MSSP) or a security information and event management (SIEM) system, the ability to capture data and the notification to the operations staff on the deployment of the system are necessary.

3.9 TYPES OF RISK

Some degree of risk is unavoidable in business. Risk is inherent in everything; however, some business processes have a higher level of risk than others, and the degree of risk varies from one activity, product or service to another. The risk practitioner should understand risk and be able to assess and respond to any risk that lies outside of the organizational risk appetite in a way that reduces it to an acceptable level. Accomplishing this task requires an ability to distinguish between inherent, residual and current risk.

3.9.1 Inherent Risk

Inherent risk is the risk level or exposure without taking into account the actions that management has taken or might take (e.g., implementing controls). For instance, a building in a given area has an inherent risk of flooding based on the characteristics of the land and nearby water.

3.9.2 Residual Risk

Residual risk is remaining risk after management has implemented a risk response, which is typically a mitigation activity but may also include risk transfer. Residual risk is calculated by subtracting the effectiveness of the risk response (typically a control) from the inherent risk. If the residual risk associated with a particular control is within the organizational risk appetite, the residual risk should be accepted. Otherwise, an additional response is needed.

Residual Risk = Inherent Risk – Cumulative Effect of Controls

3.9.3 Current Risk

When discussing risk response, the risk practitioner typically uses the term "residual risk" in a predictive sense, referring to the residual risk that will exist if a given response is implemented. This usage creates potential confusion, because the starting point for risk response is generally the risk resulting from prior actions already taken, not the strict inherent risk. Current risk is the term for risk as it exists in the moment, taking into account those actions that have already been taken but not actions that are anticipated or have been proposed.

3.10 CONTROL ACTIVITIES, OBJECTIVES, PRACTICES AND METRICS

Controls are chosen to reduce risk to acceptable levels, which requires the measurement and monitoring of risk. The metrics used in the course of risk response are significant in that they are generally carried forward to risk and control monitoring and reporting, where these same metrics are often used to measure the effectiveness of controls after they have been implemented. Organizations typically define their own metrics and thresholds for control performance, which may be compared to industry standards, benchmarks and good practices.

3.10.1 Business Processes

Risk management exists to support the organization, so control implementation should not unduly affect business operations. The risk practitioner should consider the impact of controls on the ability of the business to meet its objectives and of users to accomplish their tasks in a simple, logical manner. The risk practitioner should present this information to the risk owner in a way that allows him/her to weigh these considerations against the consequences associated with the risk.

When a business process involves sensitive information, such as personal health information (PHI), protecting the information from unauthorized disclosure, alteration or deletion is of high importance. Sensitive data may be masked or hidden from users who do not have a need to know or may be read-only for users who do not have authority to alter the data. Building these types of controls into systems and applications can be an effective form of preventive risk management. Where practical, risk responses should be designed in a manner that complies with local regulation (e.g., the US Health Insurance Portability and Accountability Act [HIPAA]), international standards (e.g., *ISO/IEC 27001—Information Security Management*), industry standards (e.g., Payment Card Industry Data Security Standard [PCI DSS]) and internal standards promulgated by the organization. Organizations that comply with standards may derive competitive advantage from certification of their compliance.

3.10.2 Information Security

One goal of risk response is ensuring that technology used in the organization is adequately protected, secure and reliable. The risk practitioner need not be an information security expert but should be sufficiently knowledgeable as to ensure that the risk assessment and response program evaluates new technology and provides advice on how to deploy and use it within acceptable risk boundaries wherever possible. Some risk-based considerations regarding the deployment of new technology include:

- Training for users and administrators
- Creation of policies and procedures
- Inclusion of such systems in backups and BCPs
- Assignment of ownership for the risk
- Consent of information owners for technology that may handle sensitive information
- Review of legal or regulatory requirements
- Assignment of responsibility for monitoring and reporting on the proper use of such technology

Legacy systems require special attention because their designs frequently fail to address current security issues. Legacy systems may also be reaching the end of their operational life and could be more susceptible to failure as a result of aging. The risk response options for legacy systems may be very limited on account of the cost involved in replacing or upgrading them. Some legacy systems may not be replaceable at all, a common occurrence in industrial environments. Under such circumstances, the risk practitioner should bear in mind that his or her role is not to reject a particular course of action but rather to work with the risk owner to create an acceptable level of risk. Compensating controls—such as additional backups, spare parts, strong network segmentation, cross-training, documentation and increased monitoring of systems performance—may be effective supplements to existing security where conventional improvements are either technically infeasible or not available.

Every system is the responsibility of a system owner. The owner of the system is usually a senior manager in the department for which the system was built. For example, the owner of financial data and the financial systems of an organization is often the chief financial officer (CFO). In such cases, the CFO is responsible for the proper use and operation of the financial system and usually must approve expenses for system implementation, changes, upgrades and removal. However, ownership does not mean literal responsibility for all actions taken. Typically, the CFO engages the IT department or an external supplier to manage and operate the financial systems on the behalf of the finance department.

It is important for the risk practitioner to understand that transference of responsibility does not mean transference of accountability, except in rare cases in which a supplier might accept full legal accountability for damages and losses. Responsibility can be delegated; accountability cannot.

An organization may have many system owners from various departments, making it difficult to manage, oversee and ensure consistent operation of the systems in a coordinated fashion. For instance, an information system that supports the sales department may operate in a considerably different cultural environment than an information system that supports the finance department. This difference in culture may affect the way the system is managed and protected. If the protection for each system is the sole responsibility of the owner of that system, then significant differences can exist in how the security and risk for each system is enforced. This is a substantial problem in modern networked organizations, because information sharing and system dependencies make it likely that a breach or vulnerability in any one system is a risk to the entire organization. For this reason, an organization should take an enterprise approach to security to ensure consistent, reliable and secure operations. Two common methods of providing enterprise consistently are change control and system authorization. Asset inventory and configuration management are also important parts of an overall information security strategy.

Change Control

Change control refers to formal review and approval of all requests to change systems or configurations by a dedicated committee. Change control boards (CCBs) are typically comprised of representatives from several business departments and are responsible for overseeing all information systems operations and approving changes to those systems, which provides a communications channel between the business units and the IT department. As changes are requested, requests are submitted for review by the CCB, which verifies that:

- The change request does not unknowingly affect risk or security.
- The change is formally requested, approved and documented.
- The change is scheduled at a time convenient for the business and IT.
- All stakeholders affected by the change are advised of the change.

The intention of the CCB model is to provide a balance between allowing needed changes to occur and preserving system reliability and stability.

System Authorization

Many organizations use a form of objective assessment and formal acceptance of risk associated with the installation and operation of information systems, culminating in the explicit authorization of a system to operate prior to it being allowed to do so. This process is known by several names, including security assessment and authorization (A&A) and certification and accreditation (C&A).

Regardless of the terminology, the purpose of system authorization is to provide an enterprise approach to the management of IT risk by ensuring that all information systems are reviewed to identify their associated risk and that this risk is either flagged or accepted prior to granting the system approval to operate. In general, this process can be broken into two parts:

1. Evaluation (assessment or certification)
2. Authorization (accreditation)

Evaluation is the process of reviewing an information system to determine the security of its design, development, testing, deployment and operations. The evaluator is a technical and process expert who examines the technical and nontechnical aspects of system operations to ensure that the risk associated with the system has been identified and to

document any mitigating controls that may be in place. To accomplish this task, the evaluator examines assessments, system documentation and the results of technical tests to verify that the system will operate within an acceptable level of risk. The culmination of the evaluation process is a report to a suitably empowered senior manager, sometimes called an accreditor or authorizing official (AO), recommending whether to authorize operation of the system. It is a good practice for evaluation to be done in parallel with the system development life cycle (SDLC) to ensure that any unacceptable risk is identified promptly and addressed to system and project management.

Authorization is the official decision by the senior manager to approve an information system for operational use. This decision is the explicit acceptance of the risk documented in the evaluator's report, and it may come with caveats. For instance, a system may be authorized to operate only under certain circumstances or in particular time frames or certain risk may be accepted only for a limited time on the basis of response activities that are already underway but have not yet been completed.

Upon receipt of authorization, the system owner may operate the system according to the restrictions and for the time period granted in the accreditation—generally one to three years, with the caveat that any substantial changes in the system or the organizational risk profile require a new evaluation and reauthorization of the system.

Asset Inventory and Documentation

An important part of managing risk is the identification and inventory of the assets of the organization. As changes are made to systems and new equipment is installed or older equipment is retired, the asset inventory must be updated to reflect what assets, systems and equipment the organization is currently using. The asset inventory should list all equipment, supplier, acquisition date, original cost, actual cost, the location, owner of the equipment and other data required for maintenance, insurance and warranty purposes.

Configuration Management

Devices and systems must typically be configured after installation to support communications, interfaces with other organizational systems and secure operations directed by policy. The goal of configuration management is to have a single approved way in which particular systems or devices intended for a particular role should be configured for use by the organization. Use of a common configuration makes it easier to deploy new systems, test patches and upgrades, identify the presence of malware, and manage the enterprise.

Configuration management often includes hardening, a process that reduces vulnerability by limiting the attack vectors that might be used as points of compromise. A hardened system is one that does not store any sensitive data that are not immediately needed to support a business operation. In addition, it has all unnecessary functionality disabled, including ports, services and protocols that are not required for the intended use of the system within the enterprise environment. Many devices and systems come with guest accounts or default passwords that should be changed or disabled as part of the hardening process.

Configurations and settings that control the operation of devices should be backed up and available in case of equipment failure. The risk practitioner should ensure that there are policies and procedures in place that apply to proper configuration management and that backups of standard configurations are available for use in recovery scenarios.

3.10.3 Third-party Management

The risk practitioner should address the risk that arises when an organization outsources business functions, IT services and data management. The ownership of the data and business processes remains with the organization that is doing the outsourcing. This creates a legal liability that may be difficult for the outsourcing organization to manage because most of the day-to-day operations, staff and procedures are outside of the organization's direct control.

When the management of data is outsourced, the outsourcing organization must ensure that the security requirements and regulations for handling the information have been written into the outsourcing agreement and are being followed. This may require the right to audit the processes of the outsource supplier or an attestation from the supplier that validates compliance. An attestation may be provided by the external auditors of the supplier or an independent reviewer.

Other issues related to outsourcing include declaring the jurisdiction of the agreement and which courts would hear any dispute related to the terms and conditions of the contract.

Some of the concerns that should be considered in relation to the risk of using an outsource supplier are:
- Hiring and training practices of the supplier
- Reporting and liaison between the outsourcing organization and supplier
- Time to respond to any incidents
- Liability for noncompliance with terms of the contract
- Nondisclosure of data or business practices
- Responding to requests from law enforcement
- Length of contract and terms for dissolution/termination of contract
- Location of data storage including backup data
- Separation between data and management of data of competing firms
- Existence and regular testing of resiliency plans (e.g., business continuity and disaster recovery)

When an organization is contracted to provide or deliver services or equipment, the risk of noncompliance with the agreement must be met through review, monitoring and enforcement of the contract terms. Any failure to meet contract terms must be identified and addressed as quickly as possible. This especially applies to the delivery of equipment that is not configured according to contractual agreements or may not provide the functionality promised during the contract negotiations and in the statement of work (SOW).

3.10.4 Data Management

Organizations regard data as being among their most valuable asset, so data should be protected at all times and in all locations, including in storage, in transit, during processing and when displayed. The risk practitioner must also be aware that data protection applies to all formats, including:
- Paper printouts and notes (e.g., reports, scratch pads)
- Magnetic media (e.g., hard drives, universal serial bus [USB] storage)
- Optical drive media
- Audio and video broadcasts
- Photos and screensavers
- Discarded material, especially anything intended for recycling

Protection of data during processing starts with reception. Input should be subject to validation checks before acceptance or processing to ensure that data are appropriately formatted and do not contain embedded commands or other content that might adversely affect automated processing systems, such as structured query language (SQL) code. The process of data validation should include:
- Range checks (e.g., allowable data values)
- Format checks (e.g., configuration of date)
- Special character checks (e.g., prevent script commands)
- Size (e.g., prevent buffer overflows [too many data] or incomplete data [not enough data])
- Likelihood (e.g., entries of correct form that are statistically unlikely and suggest an error)

Data validation may be done either with a whitelist of allowed data or a blacklist of prohibited data. A whitelist is the preferred approach in environments in which input data validation is based on static or infrequently changing values, where it is easier to anticipate what values would be reasonable than to anticipate what values might be malicious. The use of a common library for whitelisting can ensure that data validation occurs in a consistent manner across multiple applications.

Blacklists are useful in environments in which the range of valid values is extremely broad, but in which only a few known values should be prohibited (such as a dirty word search) and these are either entirely static or are updated rarely.

> **Note:** Blacklist design and maintenance can be particularly tedious whenever the same term may be expressed in multiple ways, a phenomenon known as "canonicalization." This is particularly common in web environments. For instance, the expression *www.isaca.org* can also be written as:
> - *http:/www.isaca.org*
> - *http:/www.isaca.org/*
> - *http://isaca.org*
> - *https://www.isaca.org/Pages/default.aspx*
> - *http://www%2eisaca%2eorg*
>
> Where there are many ways to input the same command, a blacklist that specifies only a subset of the possible forms may inadvertently accept an expression whose value is meant to be prohibited, leading to attacks such as directory traversal or bypassing firewall rules.

Data protection also requires assurances that changes made will not affect the integrity, precision or accuracy of the data or of data processing operations. Mechanisms that may assist in meeting this requirement include:
- Data checks and balances of input compared to output
- Checks of normal compared to abnormal levels of processing
- Anti-malware detection software
- SoD defined at every level of a system or application
- Processes requiring approval for transactions (all or based on thresholds)

One of the most important elements of data protection is control over the permissions and authorization levels of users that can access data and applications. The risk practitioner should be sure that users' permissions include the fewest privileges that they need to perform their job functions and that these privileges are granted only for the time periods in which they are necessary, known as least privilege. This is key to protecting data from improper alteration or disclosure. Maintaining a least-privilege environment requires regular review of user access permissions and the revocation of permissions whenever a user leaves the organization or changes job roles.

Data must also be protected while in storage, which includes keeping sensitive data in separate networks or on systems that are not accessible to unauthorized personnel (isolation) through some combination of network segmentation (including the use of firewalls and virtual local area networks [VLANs]), role-based access control, and physical access controls. Another tool that helps to protect data in both storage and transmission is cryptography, the use of encryption.

Identity Management
Identity management is the process of managing the identities of the entities (users, processes, etc.) that require access to information or information systems. Risk may arise from misuse of access, especially in cases where a person has a level of access that is not appropriate for their current job responsibilities.

Access should be limited to the lowest level that is required to accomplish assigned duties (least privilege) and removed when it is no longer required. Access should also be regularly reviewed to prevent accumulation of privileges over time, which may happen when a person has worked for an organization for an extended time period in different roles. Accounts that are not in regular use should be disabled to prevent their unnoticed use by a threat agent who may gain access.

Segregation of Duties
SoD is a basic internal control that prevents or detects errors and irregularities by assigning to separate individuals the responsibility for initiating and recording transactions and for the custody of assets.

SoD can be implemented by requiring one person to review and approve the work of another person. For example, a request for a large payment may need to be approved before it is processed, or a developer may need to submit code to a testing team to have it moved into production. SoD may also mean requiring two people to participate in a task simultaneously (often called dual control), a system used to control release of nuclear weapons.

The key to SoD is mutual exclusivity, meaning that a single person cannot execute both parts of the same transaction. SoD does not guarantee security, because it can be circumvented when all participants in the process agree to work together to bypass the intended effect of the control (i.e., collusion). However, because it is difficult to establish collusion, and especially to maintain it over time, SoD increases the odds that an incorrect or fraudulent transaction will be detected and addressed.

Cross-training and Job Rotation

Organizations can reduce their reliance on key staff by training multiple people with the same skills, making it possible for any of a number of individuals to step in to fill a vital role as needed. In many cases, this takes the form of cross-training, in which people on the same team are trained in one another's roles. Cross-training makes it possible for an organization to rotate people between different jobs.

Although job rotation is typically associated with business continuity and enterprise resiliency, it is also an effective security mechanism because it changes who is involved in which roles at which times, reducing the potential for collusion activities and increasing the odds that prior collusion can be detected by someone who is not part of the scheme. In environments where there is substantial concern regarding the potential for collusion, job rotation may take the form of mandatory vacations, in which people are removed from their regular duties specifically to allow independent verification. This is particularly common in financial institutions.

One risk associated with cross-training is that employees with substantially broader skill sets may be more attractive to other employers. The rotation of personnel may also result in decreased efficiency during times of transition, and employees who know that they are about to move out of jobs may be less diligent that those who expect to remain in them.

Access Control

One of the most critical areas of risk associated with information systems is the challenge of managing access control, which is commonly addressed through the concepts of identification, authentication, authorization and accountability (IAAA).

Identification

Assigning a unique identifier to every individual and process that has access to a system allows tracking and logging of the activity by the user and the possibility to investigate a problem if it were to arise. Identification is often provided through a user ID, a customer account number, an employee identification number or other unique element. Sharing of user IDs should be prohibited, and the process of issuing an employee ID should be secure and require proper authorization.

Authentication

Authentication is the process of validating an identity. After a person or process has claimed or stated his/her identity, the process of authentication verifies that the person is who they say they are. Authentication ensures that one person cannot spoof an identity or masquerade as or impersonate another user. This can also prevent the sharing of user IDs. Authentication is usually done using three methods:
- Knowledge
- Ownership (possession)
- Characteristic (biometrics)

Authentication by knowledge requires users to know a password, code phrase or other secret value to validate their identity. The risk is that knowledge is often subject to replay attacks because the item used (password) can be used for an extended period of time. This allows a person learning the password of another person to log in as that user. This is why passwords should be changed on a regular basis.

Authentication by ownership requires the use of a smart card, token, ID badge or other similar item. This requires a person to validate their identity by having an access card or providing a one-time password that is generated by the token. The problem with ownership as a means of authentication is the cost of installing the system, issuing the cards, and operating and maintaining the system. In the event that a person loses his/her card, it may be used by an imposter if the card has not been reported as lost or stolen.

Biometrics uses either physiological (e.g., fingerprints, iris scan, palm scan, etc.) or behavioral (e.g., voice print, signature dynamics) elements to authenticate a person. Biometrics is expensive, and some users find it to be intrusive and may be resistant to it.

Node authentication differs from other types of authentication because it authenticates a device or location, not an individual or process. Node authentication is used when limiting access to certain devices through the identification of Internet Protocol (IP) addresses, media access control (MAC) addresses or device serial numbers that are attempting to log on.

Strong Authentication

No single factor of authentication is strong enough on its own. All three forms can be bypassed fairly easily by a determined hacker. Therefore, it is recommended to use more than one type of authentication during the login process. This could include using a password with a smart card, or a token and biometrics as a multifactor authentication process. This is known as strong authentication.

Authorization

After a person or process has successfully logged on to a system, the system must provide that person or process the appropriate levels of authorization. Authorization refers to the privileges or permissions the person will have, including read-only, write-only, read/write, create, update, delete, full control, etc. This is where the concept of least privilege applies and ensures that employees only have the level of access they require to perform their job functions.

Temporal Isolation

A person's authorization is usually only granted for the period of time that the permissions are required. Most users can only access a building, network or application during normal working hours. Outside of those hours, access is not allowed. This prevents a user or other person from using the account at times when it should not be required.

Accountability/Auditing

The final part of the IAAA model is accountability or auditing. This action logs or records all activity on a system and indicates the user ID responsible for the activity. This underlines the need for using unique user IDs and ensuring that the logs are protected from modification or alteration. Logs are often needed for investigation and analysis and must be preserved to prevent a person from disabling the logging function or altering log data. Because logs may contain sensitive data, access to the logs should be strictly controlled.

Cryptography

Cryptography refers to the science behind encryption, a mathematical means of altering data from a readable plaintext (or cleartext) form that is easily readable into an unreadable format (ciphertext) in a manner that can be reversed by someone who has access to the appropriate key. Encryption may be used to protect the confidentiality of data by making data unreadable to anyone who is unauthorized. It protects the integrity of data by proving that the content has not been changed, the identity of the sender or both.

Data encryption provides several risk management benefits, including:
- Confidentiality
- Integrity
- Proof of origin (nonrepudiation)
- Access control
- Authentication

Encryption comes in two basic forms—symmetric and asymmetric—each of which has its own strengths and weaknesses, and these can be used in a variety of ways. The risk practitioner should be very familiar with the concepts of encryption, because it is a central component of data protection.

Symmetric Algorithms

An example of symmetric key cryptography is shown in **figure 3.6**.

Figure 3.6—Symmetric Cryptography

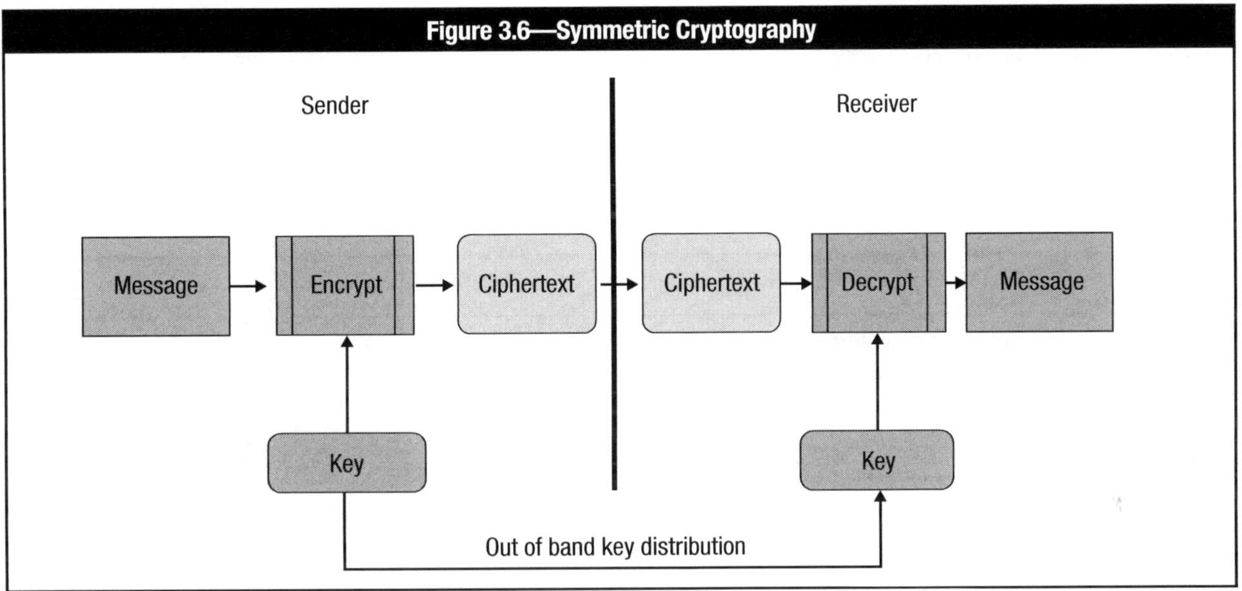

Symmetric key cryptography occurs when an encryption algorithm uses the same secret key to encrypt the plaintext into ciphertext and to decrypt the ciphertext back into plaintext. Symmetric key cryptosystems are generally less complicated than those that use asymmetric techniques and use less processing power, which makes symmetric key cryptosystems ideally suited for bulk data encryption.

The most common symmetric key cryptographic system currently in use is the Advanced Encryption Standard (AES), a public algorithm known as a block cipher because it operates on plaintext in blocks (strings or groups). AES supports keys from 128 bits to 256 bits in size and replaced both the earlier Data Encryption Standard (DES), which was a public algorithm that used 56-bit keys, and an augmented variant of DES called Triple DES (3DES) that offered somewhat stronger security.

Because the length of a key determines the number of possible mathematical combinations, shorter keys pose an inherent security problem as processing power increases. DES is no longer considered a strong cryptographic solution because its entire key space can be tested against ciphertext by large computer systems within a relatively short period of time, a technique aptly known as brute force. AES includes a number of advanced mathematical techniques that make its ciphertext stronger than what could be generated by 3DES. Nonetheless, ongoing advances in computing power virtually guarantee that any algorithm in use will eventually be rendered obsolete. Risk practitioners should also be aware that even the strongest algorithm with the most robust keys is only as secure as its key management system.

Symmetric key cryptography has two major disadvantages. One is that there is no easy way for one party to deliver keys to another with whom it would like to exchange data (particularly in e-commerce environments where customers are unknown and untrusted entities). Also, because a symmetric key is based on a mutually shared secret, there is no way to be sure which participant in a given key network originated a particular message.

Asymmetric Algorithms

Asymmetric key cryptographic systems are a relatively recent development that evolved to address the issues with symmetric key cryptosystems.

In 1976, Whitfield Diffie and Martin Hellman published the first public example of encryption based on two different keys, a technique known as the Diffie-Hellman model. In a Diffie-Hellman setup, two mathematically related keys are created—a private key and a public key. It is computationally infeasible to determine the value of one key from the other, but each has the effect, when used in an appropriate asymmetric algorithm, of creating ciphertext that only the other has the ability to return to plaintext. As a result of this relationship, two things become possible:

1. The public key may be freely distributed and used as the means of encrypting any message that should be readable only by its creator, who has sole access to the private key that allows the messages to be decrypted.
2. Because the only messages that the public key can decrypt are those that were encrypted by the private key, to which its creator has sole access, any recipient can be certain that the key creator was the true author.

Because asymmetric algorithms use a public and private key pair, they are commonly referred to as public key algorithms. Each party using public key cryptography only needs one pair of keys (the private and the corresponding public key), and because the public key can be freely distributed, public key cryptography overcomes the weakness of scalability of symmetric key cryptography.

One disadvantage to using asymmetric algorithms is they are computationally intensive and slow relative to symmetric algorithms. For that reason, asymmetric cryptography is typically used only to encrypt short messages. In fact, the most common use of asymmetric algorithms is to distribute symmetric keys that can then be used by the participants for fast, secure communication, as seen in **figure 3.7**.

Figure 3.7—Using Asymmetric Algorithms to Support Symmetric Cryptography

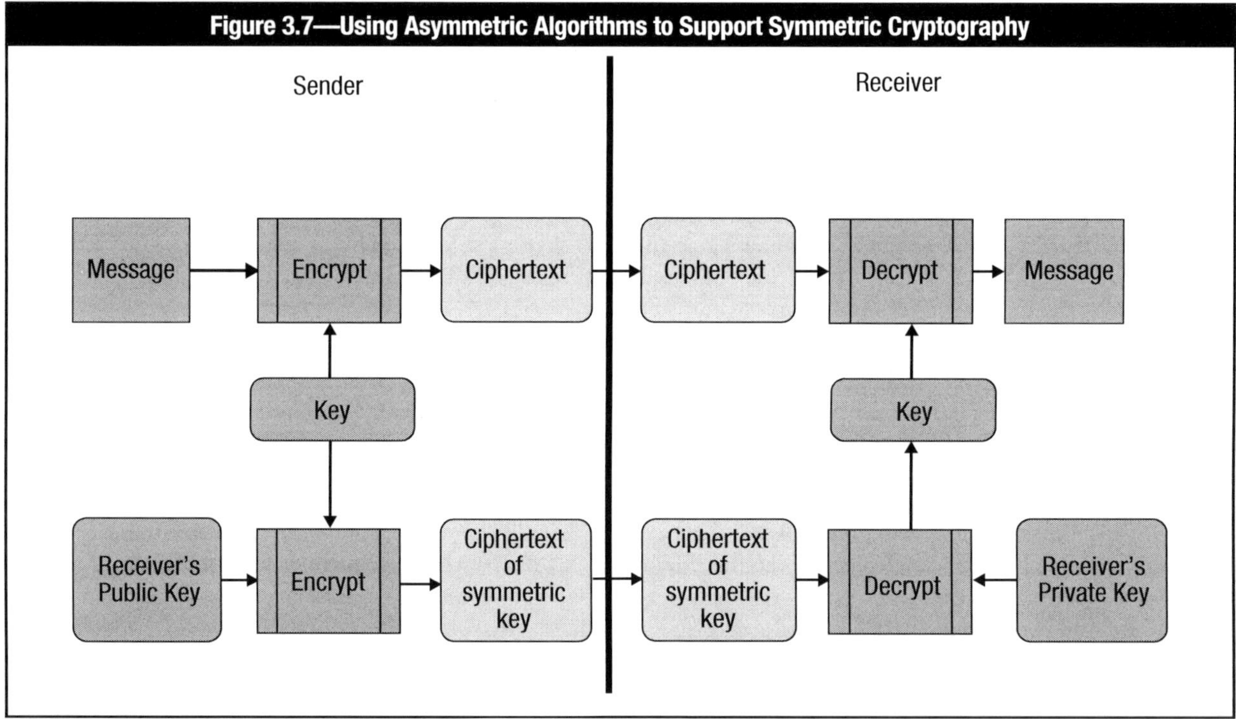

Message Integrity and Hashing Algorithms

Early computer networks used voice-grade telephone cable with limited bandwidth. The combination of slow speed and interference meant that error correcting was needed to ensure that what was received matched what was sent. Parity bits, checksums and cyclic redundancy checks (CRC) proved effective in detecting accidental errors, but these mechanisms added data to each transmission.

Hashing refers to the mathematical transformation of data using an algorithm whose result is predictable, repeatable and entirely dependent upon the content of the message and of a fixed length (regardless of the length of the original message). A hash algorithm calculates a value known as a digest, fingerprint or thumbprint from an input message whose length depends on the hash algorithm used. For example, SHA1 generates a digest of 160 bits, while SHA512 generates a digest of 512 bits. When a sender wants to ensure a message is not affected by noise or network problems, he or she can hash the message and send the digest along with the message to the receiver. The receiver can then hash the received message and verify that the resulting digest matches the digest sent with the message, as shown in **figure 3.8**.

Figure 3.8—Verifying Message Integrity Using a Hash Function

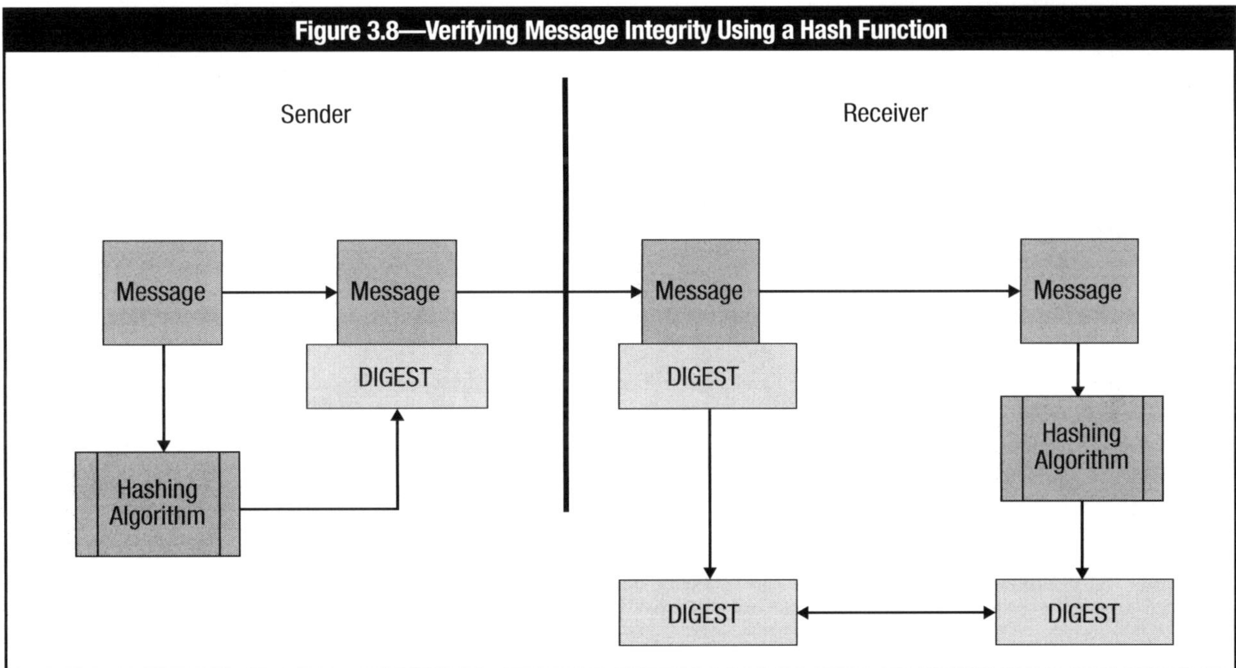

Hashing by itself provides effective protection against accidental changes in a message, such as those caused by interference. However, a malicious user who intercepts a transmission that includes a hash and changes the message likely will not be deterred by hashing, because he or she can also create a new digest to send with the replacement message.

Digital Signatures

A digital signature combines a hash function with the asymmetric encryption ability to verify the author's identity. The receiver knows that the message was not changed, because the hash of the received message is the same as the hash (digest) that was signed by the sender. The receiver also knows who the message came from, because his or her ability to decrypt the digital signature with the sender's public key means that it must have been encrypted with the sender's private key—and the sender cannot plausibly deny it, a characteristic known as "nonrepudiation" that makes digital signatures particularly valuable. **Figure 3.9** illustrates the use of digital signatures to verify integrity and proof of origin.

The risk practitioner should be clear that digitally signing a message does not make the message itself confidential. A message may be encrypted and not signed, signed and not encrypted or both signed and encrypted. Only a message that is both signed and encrypted is simultaneously afforded confidentiality, integrity and nonrepudiation.

Figure 3.9—Verifying Message Integrity and Proof of Origin Using Digital Signatures

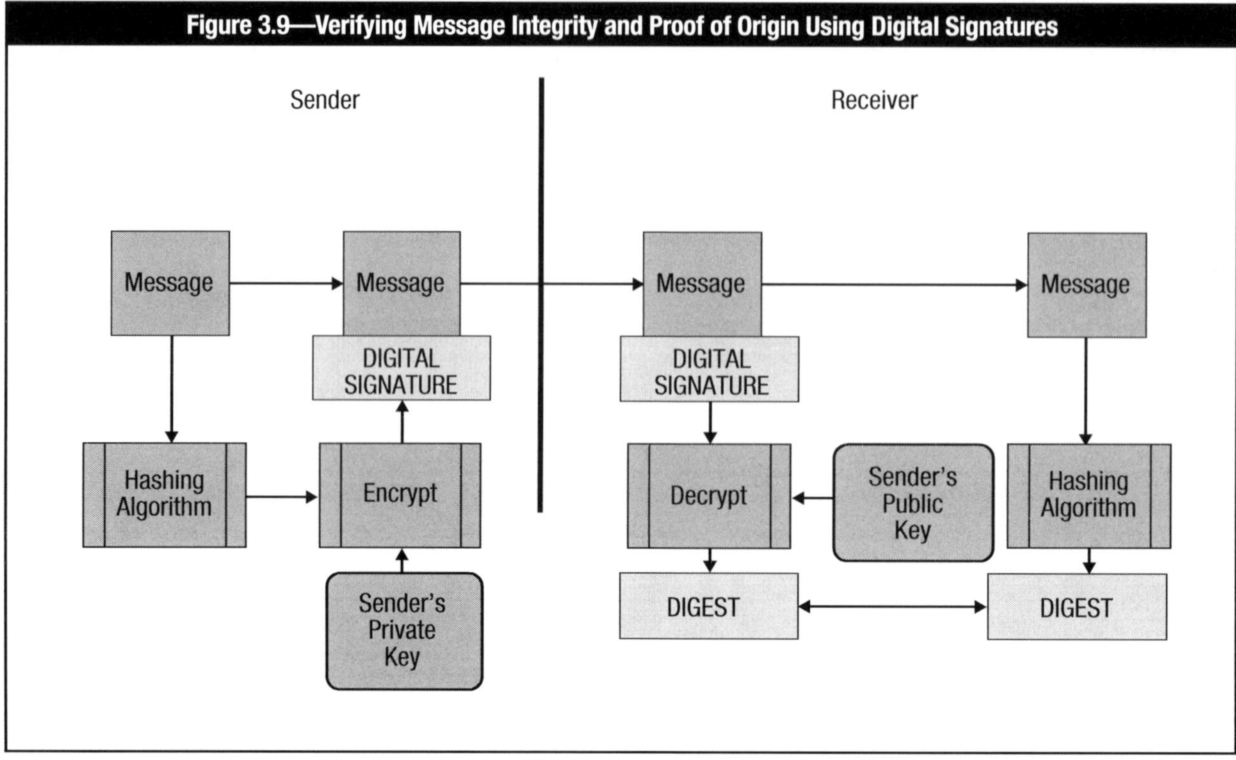

Certificates

Public key encryption ensures that only the holder of a private key can decrypt what is encrypted with the corresponding public key, but it does not offer any proof that the person who owns the public key being distributed is who he or she claims to be. The purpose of a certificate is to link a public key with a specific owner by relying on the verification of a trusted third party known as a certificate authority (CA). The CA generates a certificate on behalf of the owner of a public key that the owner can use to prove that this public key belongs to them. A recipient of a certificate created and signed by a CA can be confident that the public key in the certificate belongs to whomever is specified by the certificate, and when the recipient opens a digital signature with that public key, he/she knows that the message was signed and sent by the sender.

The format of a certificate is based on a standard called X.509, which ensures that certificates can be accessed by most browsers and systems even if they are issued by different CAs. Each certificate is valid for a defined period of time, typically one year from the date of issue. However, the owner of a certificate may cancel it at any time during that period by notifying the CA, which will put the certificate on a certificate revocation list (CRL). Requests to validate a certificate that is on a CRL result in notification that the certificate has been revoked, warning that the certificate should not be trusted as a means of verifying identity.

Public Key Infrastructure

Public key infrastructure (PKI) refers to the overall implementation of public key cryptography, including certificates and the CAs needed to issue and verify them. With a given PKI implementation, there is a top-level ("root") CA that manages all the certificates for users within the architecture. Through the root CA and any subordinate CA delegated by the root, any member of the group can gain access to the certificates of the other group members. This ensures that communication is secure and can only be accessed by the appropriate parties.

PKI implementations may be connected to one another and allow certificate verification using an external CA. When members of different PKI implementations are linked by relationships established between their respective CAs, a cross-certification agreement should be in place in order to provide a policy basis for this interconnectivity.

Additionally, the risk practitioner should ensure that mechanisms for certificate management are in place so certificates can be renewed prior to expiration to prevent an unintentional loss of access to encrypted messages and communications channels that may depend on PKI-authenticated access.

Disadvantages of Encryption

Although encryption is extremely useful and is a central component of most data protection strategies, there are drawbacks to its use. One is that encryption is a process whose processing demands increase with the length of the key and strength of the algorithm. Encryption also increases network traffic on account of the need to verify certificates, particularly in cases in which large numbers of small messages are being passed between distinct end points. The increased processing and transmission demands placed on IT infrastructures by encryption mean that it should be used only when necessary, but needs to be used for all traffic. Therefore, this requires a judgment call, which means that encryption is partially driven by policy rather than being a purely technical control. The need for policy extends to management of the CA, which can be used to issue credentials that will thereafter be generally accepted.

Encryption also presents a challenge with regards to data availability. PKI implementations are typically robust to ensure that CAs and CRLs are always available, and there are emerging mechanisms for handling matters related to encryption keys that may provide even greater speed and reliability. Nonetheless, the risk practitioner should be clear that encryption is a fail-secure control whose purposes are to guarantee confidentiality and integrity and that encrypted data may not be immediately recoverable if keys are corrupted or lost.

Summary of the Core Concepts of Cryptography

Cryptography has many moving parts that work together in a complex series of operations to accomplish various tasks. The following items summarize key information that the risk practitioner should know regarding cryptography:

- Symmetric encryption algorithms are excellent at providing confidentiality to large messages, but managing shared secret keys is difficult.
- Asymmetric algorithms are much slower than symmetric algorithms, but keys can be securely exchanged between users of asymmetric encryption using a well-established method.
- Because of the respective strengths of the two types of encryption, asymmetric algorithms are used to send symmetric keys, which are then used to encrypt messages.
- The two keys used by asymmetric algorithms (public and private) are mathematically related in a manner that lets each decrypt what the other encrypts, but it is infeasible to determine one key from the other, so public keys can be freely distributed while private keys are kept confidential.
- Hashing algorithms efficiently ensure message integrity against accidental errors by creating digests of fixed length that are mathematically linked to particular message values.
- Encrypting the hash of a message using a private key creates a digital signature that can be used to verify both message integrity and the origin of the message, which also means that the owner of the private key cannot deny sending it (nonrepudiation).
- Certificates provide assurance that the owner of a public key is who he/she claims to be. The assurance provided by a certificate depends on the degree to which the issuer of the certificate can be trusted.

3.10.5 Project, Program and Portfolio Management

Organizations rely on projects and programs to execute the decisions of management and carry out their business functions. The risk practitioner should be aware of two specific risks associated with project and program management: first, that a project may not meet its objectives, and second, that the failure of one or more projects may affect the performance of a program.

Projects are subject to numerous types of risk. Fortunately, project management methodology is well developed, and many of the common types of risk also have common responses, such as:

- Implementing a change control board to prevent scope creep
- Prioritizing critical project tasks to use resources effectively
- Reorganizing or providing additional resources to overcome bottlenecks
- Replacing or supplementing project managers who fall short of expectations
- Canceling projects whose costs or schedules are far outside of projections
- Restarting projects under more favorable conditions
- Replacing suppliers or renegotiating contracts that fail to deliver agreed-upon products or services

Organizations often associate projects with long-term programs grouped into portfolios that share common characteristics as a means of optimizing business value and creating a greater degree of oversight. Risk practitioners who identify deficiencies in project planning or execution should bring these to management's attention as soon as possible, because project trajectories become increasingly difficult to correct as they move further away from expectations.

3.10.6 The System Development Life Cycle

Risk must be evaluated and monitored at all points during the SDLC. This ensures that a system is designed, developed, tested, implemented and operated with adequate controls and protection. As the system moves through the phases of the SDLC, new risk may emerge and previous risk assessments may need to be revised. This requires the risk practitioner to work with the project team to identify and determine optimal responses to risk. This is especially important when changes are proposed to an existing application, when the risk practitioner must evaluate the potential impact of the change on the risk and security profiles of the application. Risk practitioners should be alert to whether the development team is following the standards and policies of the organization regarding system development and the implementation of controls. Failure to adhere to these requirements may represent a risk to the organization that should be escalated to management for resolution or consideration.

Additionally, the risk practitioner should be aware of the distinction between project risk, which deals with the consequences of failing to deliver the promised results on time and within budget, and risk that the deliverables may not meet the needs of the organization when they are delivered as promised. In situations that have substantial delays between the preparation of detailed requirements and delivery of final deliverables, there is substantial risk that changing circumstances will make what is delivered of minimal value even if all of the deliverables are met. Rapid prototyping and regular engagement with stakeholders for regular feedback on deliverables in progress are tools that can help ensure that projects focus on delivering what the organization needs.

3.10.7 Business Continuity and Disaster Recovery Management

Any change made to a risk environment may affect the accuracy and appropriateness of resiliency plans associated with continuity or recovery. For instance, a risk that was previously accepted by management may now require mitigation. Risk practitioners should advise teams involved in business continuity planning (BCP) or disaster recovery planning (DRP) of changes in the risk environment that may warrant review and revision of these plans.

Business impact assessments (BIAs) are valuable tools used by BCP/DRP teams and risk practitioners. The primary objective of the BIA is to identify the impact of an incident in order to understand and prioritize steps that can be taken for effective prevention or response. Each incident is a learning experience, and the risk practitioner should review BIAs and after-action reports to determine what lessons can be learned with regard both to the placement and working of controls and to incident handling procedures.

3.10.8 IT Operations Management and Acquisition

IT-related initiatives are often implemented as projects, but concerns regarding IT risk do not end with project closeout. After an IT system has been implemented, it must continue to be operated in a secure manner. The risk practitioner should take care that the agreed-upon secure standards are maintained, especially whenever changes are made to a system, network or business process.

IT departments are often the first to be notified of a system problem or potential breach, and they should be alert to system problems or abnormal conditions. Reports generated by monitoring should be provided to risk owners to alert them to any potential problem or emerging trend, and the IT department should have clear procedures in place to respond to and escalate incidents to appropriate levels based on risk.

Regular monitoring of controls should be performed to ensure that the controls are effectively mitigating risk.

The IT department is also responsible for backing up systems to enable system recovery. Encrypting backups that are stored offsite is common practice as a means of ensuring confidentiality of data, but provisions must be taken to ensure the encryption key is available when required.

The risk practitioner should review the scope of backups to ensure that it is adequate to address likely recovery needs. A comprehensive backup strategy should include:
- Files
- Transactions
- Operating systems (OSs)
- Databases
- Patches
- Configurations
- Access control lists
- Reports
- Applications

IT should ensure that purchased equipment is delivered according to specifications, including having any default settings disabled and ensuring that security features have been enabled according to instructions placed at the time of order and that the equipment is installed according to the systems architecture and design.

3.10.9 Information Systems Architecture

The architecture of an information system plays a role in determining the appropriateness of controls and other forms of risk response. For example, controls that focus on traffic at the network layer are likely to be effective in addressing threats directed against network-layer targets, but they are unlikely to be effective against application-layer attacks. Multiple controls are typically needed within an IT architecture to provide robust assurance against malicious activity.

Platforms and Operating Systems

Recently, there have been several examples of hardware infected with back doors and security vulnerabilities during the manufacturing or delivery process. These breaches have been found on network devices, point-of-sale terminals, applications and smartphones from numerous countries and vendors. Such vulnerabilities are not easy to detect. The risk practitioner should be aware of the risk of purchasing infected equipment and use trusted vendors or suppliers whenever possible. Purchasing equipment that has been tested and evaluated by an external entity using an internationally approved process, such as the Common Criteria (ISO/IEC 15408), may also provide a higher level of confidence that the equipment is secure. Once installed, the IT team must validate that maintenance hooks or back doors that vendors can use to gain access to, and monitor, the system have been secured or eliminated. All vendor-supplied default accounts and passwords should also be changed. When performing administration of hardware devices, a secure channel should be used and a form of strong authentication required.

No commercially available OS is absolutely secure; therefore, regardless of the underlying platform, it is necessary to implement strict controls over changes, patches and configurations. The systems should be hardened to disable all unnecessary services, and default accounts and passwords must be changed.

The culture of the IT operations team is also an important consideration. If the IT team is overconfident, they may not be careful enough to protect their systems and might leave them vulnerable. For example, if the UNIX support team in an organization is overconfident and does not put adequate controls into place, an attacker may be able to circumvent these weak controls and breach the security of this system. Through that breach, they might also breach the Windows system, even though the Windows support team was very careful and had implemented better controls.

Securing platforms requires a patch management process that identifies, tests and schedules the implementation of the patches. After a patch has been implemented, it should be monitored to ensure that all related systems, utilities and applications are still working correctly.

Applications

Application risk is often due to flaws or bugs in the coding of the application. This is especially true for web applications. Countermeasures to web application vulnerabilities are related to proper design, coding and testing. Resources, such as the Open Web Application Security Project (*www.owasp.org*), are available that list common web application vulnerabilities and provide direction on how to mitigate and test for such vulnerabilities.

Applications should be designed to protect information using solutions such as:
- Masking to hide sensitive data
- Menus to restrict actions that are available to the user
- Drop-down boxes to limit input options
- Range checks to ensure expected or correct data values are submitted
- Balancing to ensure proper processing of transactions
- Logs to record all activity
- Access controls to restrict user access
- Certificates to authenticate entities
- Encryption to transmit sensitive data
- Documentation to facilitate maintenance
- Coding standards to support good practice in coding

Legacy applications may have other vulnerabilities due to older standards in coding or design. These may need to be protected through the use of middleware to isolate direct access and manage data input/output, network isolation and secure communications channels.

A common problem with applications is the process of error handling. If error messages are too verbose, they might provide information to an attacker that can be used to modify the attack. Errors should be coded and provide as little information as possible to the user.

When users attempt to log in to a network, application or system, they should be required to provide a user ID and password. The use of two-factor authentication is usually desirable, especially for critical systems, and after a few invalid login attempts, the user should be denied entry. This can be accomplished by locking the user ID, disconnecting the network connection or blocking the source IP address. User IDs may remain locked for a set period of time, or they may require resetting by an administrator.

Databases

Database security is essential because databases contain critical and sensitive information that is required to support business operations.

Database security is provided through the following:
- Encryption of sensitive data in the database
- Use of database views to restrict information available to a user
- Secure protocols to communicate with the database
- Content-based access controls that restrict access to sensitive records
- Restricting administrator-level access
- Efficient indexing to enhance data retrieval
- Backups of databases (shadowing, mirroring)
- Backups of transaction journals (remote journaling)
- Referential integrity
- Entity integrity
- Validation of input
- Defined data fields (schema)

Networks

Organizations rely heavily on the ability to transfer data from one location to another. Network security includes ensuring that what is sent matches what is received (integrity), that no unauthorized party has gained access to the information in transit (confidentiality), that communications channels are available whenever needed (availability) or any combination of these factors. There are inherent trade-offs between these different aspects of network security, in part because every instance of data needing to be processed slows down throughput. Networks exist to enable the activities of the organization, so the risk practitioner should have a high-level understanding of how these different aspects of security may be attained. He/she should also understand which of these is most important to the organization in the context of a specific business process and which might be sacrificed or diminished in order to preserve the others. Like other instances of IT risk, network risk is primarily a business concern for which IT has responsibility.

Integrity of network communications may be provided through the use of parity bits, checksums, hashing and digital signatures. Integrity checks may be performed by network devices while data are in transit, by end-point recipients at the application level or both. In general, integrity checking is performed at end points when network paths are relatively short and reasonably reliable, while it may make sense to build integrity checking into routes that involve long-haul communications (such as satellite channels) to minimize the need to retransmission. Integrity solutions for in-transit communications include using IP security (IPsec) in authentication header (AH) mode and using digital signatures. Certain types of systems may also incorporate sufficient error-checking information to allow for error correction on a mathematical level. Integrity tends to be of substantial importance when dealing with data intended for use in transaction processing, including transmission of files whose presentation must be exact (such as data archives). Integrity is less important in time-sensitive processes such as voice and video traffic, where some degree of loss is accepted in favor of fewer interruptions.

Confidentiality of network data is accomplished in two ways, one of which is encryption. As with integrity checking, encryption may be provided at one or more levels. Encryption at the datalink layer is typically used in cases of open-air transmission, such as Wi-Fi Protected Access II (WPA2) for Internet protocol (IP) wireless networks or serial encryption of microwave links between adjacent devices. At the network layer, the most common mechanism for encryption is the encapsulating security payload (ESP) mode of IPsec, often through the use of virtual private network (VPN) systems that combine confidentiality with integrity checking and also authenticate user identity. Transport-layer encryption is common for web-based traffic and may be provided using Transport Layer Security (TLS) or its predecessor Secure Socket Layer (SSL), which is now considered vulnerable to compromise but is still widely in use. Secure Shell (SSH) is another mechanism for transport layer security and is commonly used for remote access to network devices. In environments in which confidentiality is paramount, application-level encryption may also be used in order to prevent access by other applications that may be running on the recipient system.

The architecture of the network is important to consider when designing or evaluating confidentiality. Isolating network segments in ways that limit access to areas of the network based on roles and responsibilities can be an effective means of preventing unauthorized access to data in transit. This is especially important for devices, such as point-of-sale terminals, that handle sensitive data and for channels used to manage network devices. Administration of network devices should only be performed by authorized personnel and only through a change control and quality assurance process. Even a small error by a network administrator can lead to serious consequences. When performing remote administration using a protocol such as Simple Network Management Protocol (SNMP) versions 1 or 2, the connection should be over a secure encrypted connection (such as layer 2 tunneling protocol running over IPsec) because those protocols are not secure. Supervisory control and data acquisition (SCADA) systems and industrial control systems (ICS) used to monitor and control industrial machinery such as power plants, physical/environmental controls and pipelines are particularly dependent upon network segmentation because the real-time nature of their operations tends to make encryption infeasible as a solution even when the systems technically support it (and many of these systems do not support it).

Network segmentation is often provided through the use of firewalls and gateways deployed in layers, so that an attacker must overcome a series of distinct obstacles in order to successfully exploit the network. Devices used for layered defense include intrusion detection systems (IDSs) and intrusion prevention systems (IPSs), virtual local-area networks (VLANs) and bastion hosts. When an organization finds it beneficial to provide outside access to products, data or services, the best approach is often to design a demilitarized zone (DMZ), which is a screened (firewalled) network segment that acts as a buffer zone between a trusted and untrusted network. A DMZ should only provide the minimal levels of access required. No sensitive data should be stored in a DMZ, and all devices in the DMZ should be hardened with limited functionality. Where highly reliable ongoing communications are needed between trusted organizations, the solution may be to build a private network, called an extranet, that connects internal networks on the basis of secure authentication. It is good practice to provide different connections for different functions that allow tailored security measures for every function.

Network availability is provided by creating multiple paths for data to travel so that communication remains possible even in the event of a partial network failure. Specific mechanisms include alternate routing, redundancy of cable and network devices, and load balancing. Successful communication within a network requires participation by multiple devices operating at different levels and serving specialized functions, and risk practitioners should understand the fundamentals of network communications in order to reasonably review and discuss network availability.

3.11 SYSTEMS CONTROL DESIGN AND IMPLEMENTATION

Systems should be designed with security in mind, and the development process should ensure that security controls are built into the system and tested prior to deployment. Additionally, each component of a system should be tested as it is developed or acquired. Changeover from one system to another includes changes to substantial components of system or movement from one distinct version to another. This process must be deliberate, planned and structured in a manner that presents an acceptable level of risk. Clear plans for rollback and a formal closeout process for projects that deliver systems are also important in the context of overall risk management.

3.11.1 Testing

Organizations should test everything that they produce, whether manufactured on a physical or logical basis, including software and application code. The risk of releasing a new or modified program that fails to meet business requirements or that results in a security breach is a significant factor in the risk management of the organization. Testing is an opportunity to uncover flaws early enough not only to prevent that failure, but to do so in a cost-effective manner. A comprehensive testing program includes testing at the unit/component, integration/system, and end-user levels, with good practices based on standards used at each level and results of all tests documented.

Testing may be progressive or regressive. Progressive testing begins with expectations and looks for flaws, while regressive testing works backward from known problems to identify causes. Both forms of testing are commonly used throughout the testing process. The risk practitioner may find regression testing to be useful in determining whether incidents have root causes in policy or standards.

Good Practices for Testing

Good practices for testing include considerations for data, version control and code. These are discussed in the following sections.

Data

The validity of testing depends in large part on the appropriateness of test data. All sensitive data elements should be displayed in a manner that makes it unreadable (obfuscated), such as masking sensitive information by displaying it as special characters or altering the data itself. Test data should be complete and allow the testing of all possible process functions and error handling, which includes "fuzzing"—testing the limit of the acceptable range of values and values beyond the allowable range in order to verify the functionality of input validation and process integrity controls.

Although testing simulates production use, the behavior of a system being tested is typically not fully predictable. Using distinct test data rather than production data is preferable in order to prevent the disclosure of sensitive data to unauthorized personnel. Production data should be used as test data only in exceptional cases and with specific management approval.

Environmental Separation

Development and production areas should be separated to prevent the potential for intentional or unintentional cross-population of data or application code outside of the approval process. Depending on the organization, this separation may be done at the network level, include physically separate workspaces, or even require entirely distinct teams of developers working on testing and production. The risk practitioner should understand the separation model used by the organization and the rationale for managing separation in that fashion.

Version Control

There is always risk that a change to a system will overwrite or bypass functionality previously implemented or changed in a system. Version control refers to the assignment of specific version numbers of each revision of a system, making it possible for risk practitioners and engineers to distinguish between versions. Version control tracks the current version of the software and checks for previous changes when a new change is implemented and can be used to ensure that the correct version of the source code of the software is being used in compiling the object code for production use. The risk practitioner should take care to ensure that reports from code reviews and other testing apply to the correct versions of components and systems.

Code

Part of the process of moving a new or modified system or application into production is to lock down the code in order to ensure that it cannot be inadvertently modified after it has been approved for final testing. Some organizations call this final version gold code or locked code.

Unit Testing and Code Review

Systems function on the basis of the combined effort of their components, and as the components are integrated, it can be harder to find problems. Developers should be held accountable for following organizational coding standards, including proper use of comments, and these standards should reflect good practices of the industry wherever compatible with organizational goals. Additionally, all new or changed source code should be reviewed by a third party to validate compliance prior to production use. Third-party review is valuable because it can detect both unauthorized changes made by the programmer and implementations of error handing, input validation or documentation that may be inadequate. The process of third-party review should be done by an impartial, knowledgeable party and should be based on standards, not programmer preferences.

Unit testing of each component of a system is important beyond the software environment. Testing should be done by individuals not involved in the creation of the deliverables being tested. Where detailed engineering information is available, it may be possible to perform this testing from a vulnerability assessment perspective. Commercial software and vendor products typically do not provide sufficient insight into their fundamental operations to allow for this informed perspective, making a review a more subjective assessment based on the penetration-testing model.

Unit testing performed with knowledge of the code is typically referred to as white box testing, while the term black box testing is used to refer to testing in which the behavior ofthe system must be mapped out with no knowledge of how the code is written.

It is unlikely that the risk practitioner will be expected to conduct unit testing or code reviews, but he or she should be aware of the concepts and limitations of each approach, as well as ensuring that standards are in place and are used as the basis for any testing that is performed.

Integration Testing/System Testing

After the individual components of a system have been tested, the entire system should be tested to assess how the components work together with their interfaces and their overall operation. Integration testing is distinct from component-level testing because it shows the interactions of the system with upstream and downstream peers. Unit testing and initial integration testing is often performed in a separate area from the final system testing. This separates the developer from the final testing and acceptance process and helps ensure that the components being tested are not being modified by the developer during the testing process. It also provides a level of assurance that where system problems arise, they are integration problems and not problems with the individual components.

One purpose of system testing is to assess the security functions designed into the system and ensure that it has been built according to the design (taking into account any deviations that have been approved and documented). **Figure 3.10** describes some of the different testing options available at the system level.

Figure 3.10—Options for System Testing

Test	Description
Recovery	Checks the system's ability to recover after a software or hardware failure
Security	Verifies that the modified/new system includes provisions for appropriate access controls and does not introduce any security holes that may compromise other systems
Stress	Determines the maximum number of concurrent users/services the application can process by increasing the number of users/services on an incremental basis
Volume	Determines the maximum volume of records (data) that the application can process by increasing the volume on an incremental basis
Stress/volume	A hybrid approach that uses large quantities of data to evaluate performance during peak hours
Performance	Compares the performance of the subject system to similar systems using well-defined benchmarks

User Acceptance Testing

Prior to implementation of a new or modified system, the system should be approved by its intended users through a rigorous process of UAT. A project that delivers what it promised, but not what the organization needs, cannot be called successful beyond a semantic level. UAT may highlight problems with functionality, training or process flow not detected earlier in the process. For this reason, IT project management is increasingly incorporating UAT into the development process on an iterative basis, so that key functionality can be assessed early enough that changes may be considered and approved without negatively affecting the project schedule.

The purpose of UAT is to verify that the system meets user requirements and expectations, not whether it meets the stated design. A failure in UAT suggests flaws in the organization's processes for needs analysis and requirements definition.

Quality Assurance

Quality assurance (QA) is a planned and systematic plan of all actions necessary to provide adequate confidence that an item or product conforms to established technical requirements. It has been developed and documented according to organizational standards and good practice. A mature QA program is a planned and systematic plan of actions needed to provide confidence that a deliverable conforms to established requirements. QA addresses whether the project delivered what it promised in the stated design, not whether that design actually meets user requirements and expectations. A failure in QA suggests flaws in the organization's processes for development and execution.

3.11.2 Changeover (Go-live) Techniques

After final tests have been conducted, implementation is scheduled. There are three common changeover methods used to transition from one system or version to another:

- Parallel
- Phased
- Abrupt

Parallel Changeover

A parallel changeover means operating both the new system and the old system simultaneously. Parallel processing allows the project team greater latitude to test the reliability and performance of the new system and ensure that it is working correctly before removing the old system from service, which minimizes the risk of a failed changeover to the new systems and allows staff to train on and become familiar with the new system before it is in full production.

Disadvantages of a parallel changeover include the cost of maintaining both systems at the same time and ensuring that the data are consistent between both systems (because both are operational).

Figure 3.11 shows the principles of a parallel changeover.

Figure 3.11—Parallel Changeover

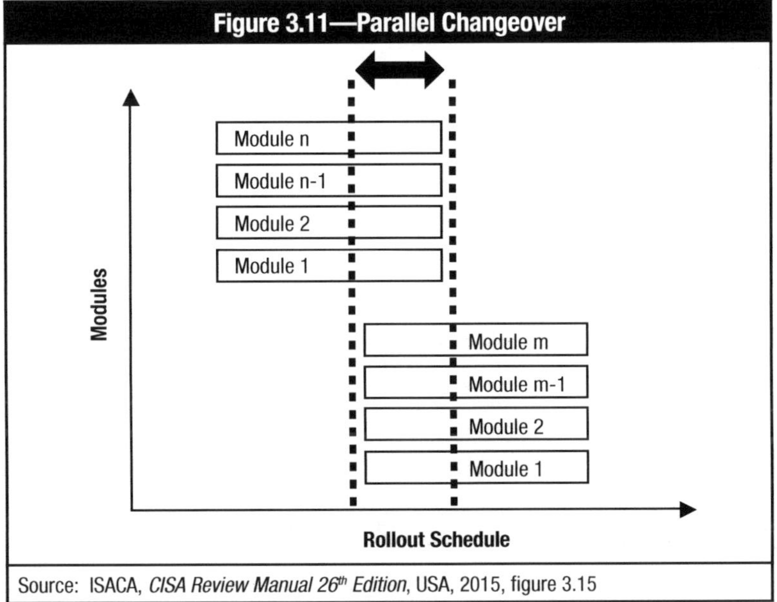

Source: ISACA, *CISA Review Manual 26th Edition*, USA, 2015, figure 3.15

Phased Changeover

A phased changeover is conducted by replacing individual components or modules of the old system with new or modified components. This reduces the risk by gradually rolling out the new modules without impacting the entire system. Depending on the type of system, this type of changeover is not always possible. Some of risk factors that may exist in a phased changeover include:

- IT resource challenges arising from having to maintain two unique environments with distinct hardware, OSs, databases and code, etc.
- Operational resource challenges arising from having to maintain two different sets of user guides, procedures and policies, definitions of system terms, etc.
- Challenges maintaining consistency of data on multiple systems or locations
- Extension of the project life cycle to cover two systems
- Change management for requirements and customizations to maintain ongoing support of the older system

Figure 3.12 shows the principles of a phased changeover.

Figure 3.12—Phased Changeover

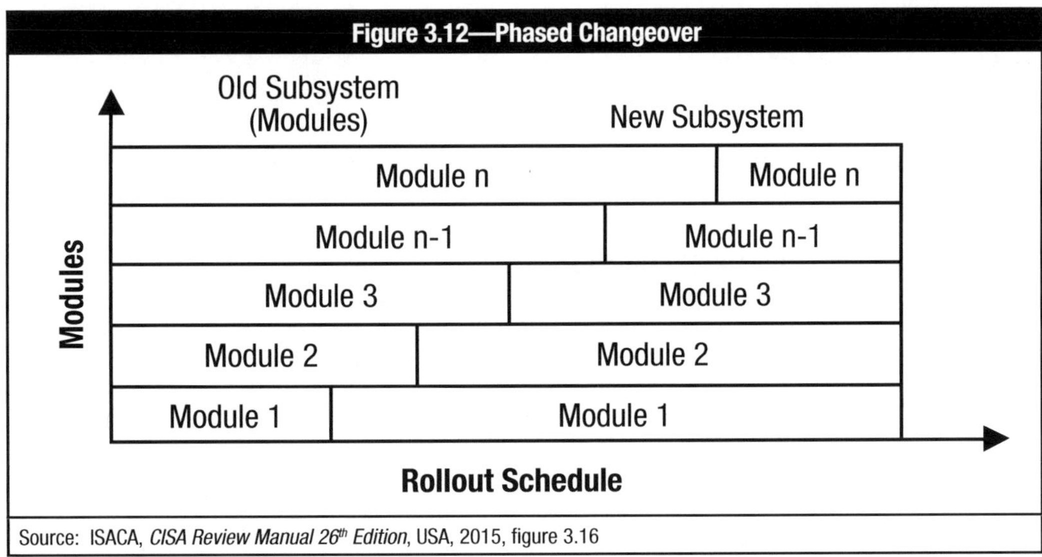

Source: ISACA, *CISA Review Manual 26th Edition*, USA, 2015, figure 3.16

Abrupt Changeover

An abrupt changeover means single-instant movement from the old system to the new system, with the old system immediately taken offline. Abrupt changeover is the riskiest of the changeover processes because of the potential for lost opportunities in business processing if it becomes necessary to roll back to the earlier system (assuming rollback is possible at all). Abrupt changeover is most prevalent in cases in which rollback is relatively assured or the impact of lost processing is minor.

The steps in an abrupt changeover are:
1. Convert files and programs.
2. Perform test runs on the test bed.
3. Install new hardware, OS, application system and migrated data.
4. Train employees or users in groups.
5. Schedule operations and test runs for go-live or changeover.

Figure 3.13 demonstrates the principles of an abrupt changeover.

Figure 3.13—Abrupt Changeover

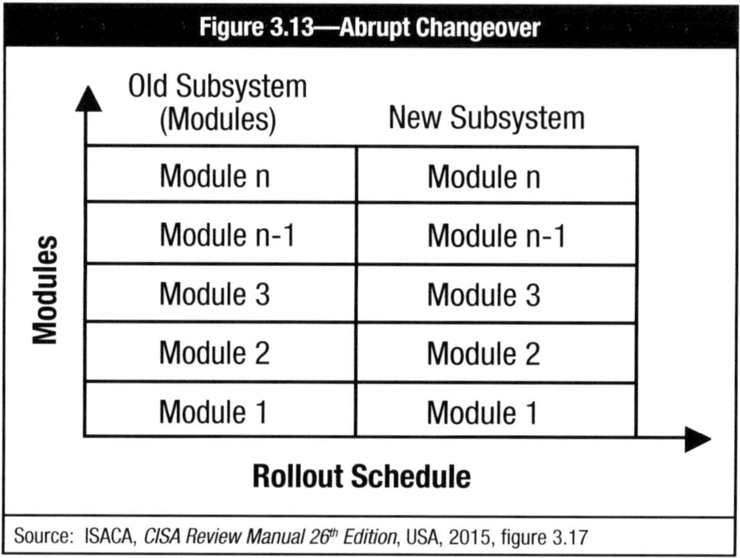

Source: ISACA, *CISA Review Manual 26th Edition*, USA, 2015, figure 3.17

Risk associated with an abrupt changeover includes:
- Asset safeguarding
- Data integrity
- System effectiveness
- System efficiency
- Change management challenges (depending on the configuration items considered)
- Duplicate or missing records (duplicate or erroneous records may exist if data cleansing is not done correctly)

3.11.3 Challenges Related to Data Migration

Migrating from one system or making fundamental modifications to an existing system may require the migration or conversion of data, which poses a risk to data integrity and availability. The risk practitioner should assess the process used for data migration. Note that there is risk associated to the process. Some considerations for data conversion are listed in **figure 3.14**.

Figure 3.14—Data Conversion Key Considerations	
Consideration	**Guidelines for the Risk Practitioner**
Completeness of data conversion	Verify that the total number of records from the source database is transferred to the new database (assuming the number of fields is the same).
Data integrity	Verify that the data are not altered manually, mechanically or electronically by a person, program or substitution or by overwriting in the new system. Integrity problems also include transposition and transcription errors or problems transferring particular records, fields, files and libraries.
Storage and security of data under conversion	Verify that data are backed up before conversion for future reference or any emergency that may arise out of data conversion program management and that controls are in place to safeguard data while in storage or transit. **Note:** Unauthorized or unnecessary copies may lead to misuse, abuse or theft of data from the system.
Data consistency	Verify that the field/record called for from the new application is consistent with that of the original application, which ensures consistency during repeat testing.
Business continuity	Verify that the application supports newer records as added or appended, which is important for ensuring seamless business continuity.

3.11.4 Fallback (Rollback)

When implementing a new system or making changes to an existing system, there is risk that the change will not work as anticipated. Failure might be due to technical problems, software problems or other unexpected challenges. Regardless of the cause, the project team should have a fallback plan so that it is possible to return to the prior system or configuration. Good practices dictate that tools and procedures required to roll back changes should be available prior to beginning the production changeover.

3.11.5 Postimplementation Review

Projects provide opportunities to learn lessons to enable more effective results for future projects. A postimplementation review is intended to document these lessons, including whether the project was properly designed, developed, implemented and managed and whether appropriate controls were built into the system. It should take into account both the technical details and the process that was followed in the course of the project, including the following:

- Adequacy of the system:
  - Does the system meet user requirements and business objectives?
  - Have controls been adequately defined and implemented?
- Projected cost versus benefits or ROI measurements
- Recommendations that address any system inadequacies and deficiencies
- Plan for implementing any recommendations
- Assessment of the development project process:
  - Were the chosen methodologies, standards and techniques followed?
  - Were appropriate project management techniques used?
  - Is the risk of operating the system within acceptable risk levels?

Not all lessons associated with a given project may be immediately evident upon completion. It is good practice for the project development team and a selection of end users to perform a second, joint review after the project has been completed and the system has been in production for a sufficient time period to assess its effectiveness and value to the organization.

3.11.6 Project Closeout

Projects are distinguished from programs by their finite life span. A project may last months or years, but there should always be a point at which any deliverables are fully transitioned to the users and/or system support staff and the project is closed.

Closing a project is a formal process that focuses on capturing lessons learned for future use. **Figure 3.15** summarizes five steps that should be included in the closing of a project.

Figure 3.15—Project Closeout Steps

Step	Action
1	Assign responsibility for any outstanding issues to specific individuals and identify the related budget for addressing these issues (if applicable).
2	Assign custody of contracts, and either archive documentation or pass it on to those who need it.
3	Conduct a postimplementation review with the project team, development team, users and other stakeholders to identify lessons learned that can be applied to future projects. Include the following information: • Content-related criteria, such as: – Fulfillment of deliverable targets and any additional objectives – Attainment of project-related incentives – Adherence to the schedule and costs • Process-related criteria, such as: – Team dynamics and internal communication – Relationships between the project team and external stakeholders
4	Document any risk that was identified in the course of the project, including risk that may be associated with proper use of the deliverables, and update the risk register.
5	Complete a second postimplementation review after the project deliverables have been completed for long enough to realize the true business benefits and costs, and use the results of this review to measure the project's overall success and impact on the business.

Note: Project closeout should include verification that the project sponsor is satisfied with the deliverables.

3.12 IMPACT OF EMERGING TECHNOLOGIES ON DESIGN AND IMPLEMENTATION OF CONTROLS

Many organizations take a proactive approach to the use of new technologies, which requires the risk practitioner to be alert to the development of new technologies and to assess the risk of their incorporation into the enterprise. In some cases, eagerness to bring new technology into use may get ahead of policies put in place to govern the process of adopting that technology. The risk practitioner may find it beneficial to use discovery scanners to detect unauthorized implementations of new devices or technologies that have not yet been reviewed and approved for use. Staff should be aware of the processes required to submit a new technology for consideration for use, and any exceptions to full evaluation should be documented and approved by management.

3.13 CONTROL OWNERSHIP

Risk accountability and ownership are not always defined or clearly stated. All risk identified in the risk register should have clearly specified owners at a management level empowered to make decisions on behalf of the organization. The risk practitioner should communicate with risk owners to ensure that they are aware of implemented risk responses and those that are pending implementation as well as the level of risk that the organization is currently facing. Unwillingness of a risk owner to accept current risk mandates the adoption of a new risk response, continued through iterations until the risk reaches an acceptable level, at which point the risk owner should formally accept the risk. Acceptance of residual risk should also include accountable ownership of the controls that produce the appropriate mitigations to ensure that these remain in effect after acceptance.

3.14 RISK MANAGEMENT PROCEDURES AND DOCUMENTATION

The risk practitioner should work to develop risk management solutions that work effectively and efficiently. Risk management is often accomplished through the use of a combination of administrative, technical and physical controls. The risk practitioner should play a key role in ensuring that controls are set up and operated correctly and are maintained, evaluated and reported to management on a regular basis. Control management procedures include:

- Proper installation
- Creation of policies and procedures to support operations
- Implementation of change management procedures to ensure correct configuration
- Training of staff to monitor, manage and review controls
- Assignment of responsibility for monitoring and investigation (which may be distinct)
- Creation of a schedule for review and reporting

As risk responses are implemented, the risk practitioner should ensure that each project either has been implemented according to the intent and design of the project architects or that any changes have been reviewed to ensure that they did not erode or diminish effectiveness. Each control should have its own documentation that includes justification for its implementation, its accountable owner, and its review and reporting schedules.

Similarly, as projects are completed and residual levels of risk are accepted by risk owners, the risk register should be updated to reflect the changes. The risk register should also show the progress of testing and the attainment of milestones during the progress of a mitigation project. By keeping the risk register accurate and up-to-date, the risk practitioner ensures that it is consistently available as a resource for risk management activities.

3.15 SUMMARY

The organization should be aware of its risk, know the assessed level of the risk and take steps to document its risk in the risk register. Based on the risk assessment, the risk response options are examined and a suitable risk response is selected. Projects and programs are developed to implement the risk response, and the results are tracked in the risk register. Controls are designed and implemented, and the level of risk is reduced to a level that the management of the organization can accept. Changes in risk and the effectiveness of controls should continue to be monitored, and this information should be reported to management.

ENDNOTES

[1] International Organization for Standardization (ISO)/International Electrotechnical Commission (IEC); *ISO/IEC 27001:2013 Information technology—Security techniques—Information security management systems—Requirements*, Switzerland, 2013

Page intentionally left blank

Chapter 4: Risk and Control Monitoring and Reporting

Section One: Overview

Section Two: Contents

Section One: Overview

DOMAIN DEFINITION

Continuously monitor and report on IT risk and controls to relevant stakeholders to ensure the continued efficiency and effectiveness of the IT risk management strategy and its alignment to business objectives.

LEARNING OBJECTIVES

The objective of this domain is to ensure that the CRISC candidate has the knowledge necessary to:

- Differentiate between key risk indicators (KRIs) and key performance indicators (KPIs)
- Describe data extraction, aggregation and analysis tools and techniques
- Compare different monitoring tools and techniques
- Describe various testing and assessment tools and techniques

CRISC EXAM REFERENCE

This domain represents 22 percent of the CRISC exam (approximately 33 questions).

TASK AND KNOWLEDGE STATEMENTS

TASKS

There are seven tasks within this domain that a CRISC candidate must know how to perform. These relate to IT risk identification.

T4.1 Define and establish key risk indicators (KRIs) and thresholds based on available data, to enable monitoring of changes in risk.
T4.2 Monitor and analyze key risk indicators (KRIs) to identify changes or trends in the IT risk profile.
T4.3 Report on changes or trends related to the IT risk profile to assist management and relevant stakeholders in decision making.
T4.4 Facilitate the identification of metrics and key performance indicators (KPIs) to enable the measurement of control performance.
T4.5 Monitor and analyze key performance indicators (KPIs) to identify changes or trends related to the control environment and determine the efficiency and effectiveness of controls.
T4.6 Review the results of control assessments to determine the effectiveness of the control environment.
T4.7 Report on the performance of, changes to, or trends in the overall risk profile and control environment to relevant stakeholders to enable decision making.

KNOWLEDGE STATEMENTS

The CRISC candidate should be familiar with the task statements relevant to each domain in the CRISC job practice. The tasks are supported by 41 knowledge statements that delineate each of the areas in which the risk practitioner must have a good understanding in order to perform the tasks. Many knowledge statements support tasks that cross domains.

The CRISC candidate should have knowledge of:

1. Laws, regulations, standards and compliance requirements
2. Industry trends and emerging technologies
3. Enterprise systems architecture (e.g., platforms, networks, applications, databases and operating systems)
4. Business goals and objectives
5. Contractual requirements with customers and third-party service providers
6. Threats and vulnerabilities related to:
    - 6.1 Business processes and initiatives
    - 6.2 Third-party management
    - 6.3 Data management
    - 6.4 Hardware, software and appliances

   6.5 The system development life cycle (SDLC)
   6.6 Project and program management
   6.7 Business continuity and disaster recovery management (DRM)
   6.8 Management of IT operations
   6.9 Emerging technologies
7. Methods to identify risk
8. Risk scenario development tools and techniques
9. Risk identification and classification standards, and frameworks
10. Risk events/incident concepts (e.g., contributing conditions, lessons learned, loss result)
11. Elements of a risk register
12. Risk appetite and tolerance
13. Risk analysis methodologies (quantitative and qualitative)
14. Organizational structures
15. Organizational culture, ethics and behavior
16. Organizational assets (e.g., people, technology, data, trademarks, intellectual property) and business processes, including enterprise risk management (ERM)
17. Organizational policies and standards
18. Business process review tools and techniques
19. Analysis techniques (e.g., root cause, gap, cost-benefit, return on investment [ROI])
20. Capability assessment models and improvement techniques and strategies
21. Data analysis, validation and aggregation techniques (e.g., trend analysis, modeling)
22. Data collection and extraction tools and techniques
23. Principles of risk and control ownership
24. Characteristics of inherent and residual risk
25. Exception management practices
26. Risk assessment standards, frameworks and techniques
27. Risk response options (i.e., accept, mitigate, avoid, transfer) and criteria for selection
28. Information security concepts and principles, including confidentiality, integrity and availability of information
29. Systems control design and implementation, including testing methodologies and practices
30. The impact of emerging technologies on design and implementation of controls
31. Requirements, principles, and practices for educating and training on risk and control activities
32. Key risk indicators (KRIs)
33. Risk monitoring standards and frameworks
34. Risk monitoring tools and techniques
35. Risk reporting tools and techniques
36. IT risk management best practices
37. Key performance indicator (KPIs)
38. Control types, standards, and frameworks
39. Control monitoring and reporting tools and techniques
40. Control assessment types (e.g., self-assessments, audits, vulnerability assessments, penetration tests, third-party assurance)
41. Control activities, objectives, practices and metrics related to:
   41.1 Business processes
   41.2 Information security, including technology certification and accreditation practices
   41.3 Third-party management, including service delivery
   41.4 Data management
   41.5 The system development life cycle (SDLC)
   41.6 Project and program management
   41.7 Business continuity and disaster recovery management (DRM)
   41.8 IT operations management
   41.9 The information systems architecture (e.g., platforms, networks, applications, databases and operating systems)

SELF-ASSESSMENT QUESTIONS

CRISC self-assessment questions support the content in this manual and provide an understanding of the type and structure of questions that have typically appeared on the exam. Questions are written in a multiple-choice format and designed for one best answer. Each question has a stem (question) and four options (answer choices). The stem may be written in the form of a question or an incomplete statement. In some instances, a scenario or a description problem may also be included. These questions normally include a description of a situation and require the candidate to answer two or more questions based on the information provided. Many times a question will require the candidate to choose the **MOST** likely or **BEST** answer among the options provided.

In each case, the candidate must read the question carefully, eliminate known incorrect answers and then make the best choice possible. Knowing the format in which questions are asked, and how to study and gain knowledge of what will be tested, will help the candidate correctly answer the questions.

4-1 The **MOST** important reason to maintain key risk indicators (KRIs) is that:

A. complex metrics require fine-tuning.
B. threats and vulnerabilities change over time.
C. risk reports need to be timely.
D. they help to avoid risk.

4-2 Which of the following choices is the **BEST** measure of the operational effectiveness of risk management process capabilities?

A. Key performance indicators (KPIs)
B. Key risk indicators (KRIs)
C. Base practices
D. Metric thresholds

4-3 During a data extraction process, the total number of transactions per year was forecasted by multiplying the monthly average by twelve. This is considered:

A. a controls total.
B. simplistic and ineffective.
C. a duplicates test.
D. a reasonableness test.

4-4 The **BEST** test for confirming the effectiveness of the system access management process is to map:

A. access requests to user accounts.
B. user accounts to access requests.
C. user accounts to human resources (HR) records.
D. the vendor database to user accounts.

4-5 Which of the following choices provides the **BEST** assurance that a firewall is configured in compliance with an enterprise's security policy?

A. Review the actual procedures.
B. Interview the firewall administrator.
C. Review the parameter settings.
D. Review the device's log file for recent attacks.

4-6 One way to verify control effectiveness is by determining:

A. its reliability.
B. whether it is preventive or detective.
C. the capability of providing notification of failure.
D. the test results of intended objectives.

4-7 Tools that correlate information from multiple systems to improve trend analysis are **MOST** likely to be applied to:

A. transaction data.
B. configuration settings.
C. system changes.
D. process integrity.

4-8 Which of the following methods is the **MOST** effective way to ensure that outsourced service providers comply with the enterprise's information security policy?

A. Periodic audits
B. Security awareness training
C. Penetration testing
D. Service level monitoring

ANSWERS TO SELF-ASSESSMENT QUESTIONS

Correct answers are shown in **bold**.

4-1 A. While most key risk indicator (KRI) metrics need to be optimized in respect to their sensitivity, the most important objective of KRI maintenance is to ensure that KRIs continue to effectively capture the changes in threats and vulnerabilities over time.
B. Threats and vulnerabilities change over time, and KRI maintenance ensures that KRIs continue to effectively capture these changes.
C. Risk reporting timeliness is a business requirement but is not a driver for KRI maintenance.
D. Risk avoidance is one possible risk response. Risk responses are based on KRI reporting.

4-2 **A. Key performance indicators (KPIs) are assessment indicators that support judgment regarding the performance of a specific process.**
B. Key risk indicators (KRIs) only provide insights into potential risk that may exist or be realized within a concept or capability that they monitor, not necessarily at the process level.
C. Base practices are activities that, when consistently performed, contribute to achieving a specific process purpose. However, base practices need to be complemented by work products to provide reliable evidence about the performance of a specific process.
D. Metric thresholds are decision or action points that are enacted when a KPI or KRI reports a specific value or set of values.

4-3 A. The described test does not ensure that all transactions have been extracted.
B. While simplistic, the reasonableness test is a valid foundation for more elaborate data validation tests.
C. The described test does not identify duplicate transactions.
D. Reasonableness tests make certain assumptions about the information as the basis for more elaborate data validation tests.

4-4 A. Mapping access requests to user accounts confirms that all access requests have been processed; however, the test does not consider user accounts that have been established without the supporting access request.
B. Mapping user accounts to access requests confirms that all existing accounts have been approved.
C. Mapping user accounts to human resources (HR) records confirms whether user accounts are uniquely tied to employees.
D. Mapping vendor records to user accounts may confirm valid accounts on an e-commerce application but is flawed because it does not consider user accounts that have been established without the supporting access request.

4-5 A. While procedures may provide a good understanding of how the firewall is supposed to be managed, they do not reliably confirm that the firewall configuration complies with the enterprise's security policy.
B. While interviewing the firewall administrator may provide a good process overview, it does not reliably confirm that the firewall configuration complies with the enterprise's security policy.
C. A review of the parameter settings provides a good basis for comparison of the actual configuration to the security policy and reliable audit evidence documentation.
D. While reviewing the device's log file for recent attacks may provide indirect evidence about the fact that logging is enabled, it does not reliably confirm that the firewall configuration complies with the enterprise's security policy.

4-6 A. Reliability is not an indication of control strength; weak controls can be highly reliable, even if they do not meet the control objective.
B. The type of control (preventive or detective) does not help determine control effectiveness.
C. Notification of failure does not determine control strength.
D. Control effectiveness requires a process to verify that the control process worked as intended and meets the intended control objectives.

4-7 **A. Transaction data tends to be difficult to analyze at a personal level for trends across multiple systems because of sheer volume and level of detail. Therefore, correlation engines are often used to analyze the data to see trends that otherwise might not be apparent.**
B. Configuration settings are generally compared against predefined values rather than between systems, and trends in configuration are rarely of value.
C. System changes are compared from a previous state to the current state, and only in certain limited circumstances would it be reasonable to subject system changes to trend analysis.
D. Process integrity depends on system integrity and is not typically associated with trend analysis.

4-8 **A. Regular audits can identify gaps in information security compliance.**
B. Training can increase user awareness of the information security policy but is not more effective than auditing.
C. Penetration testing can identify security vulnerability but cannot ensure information compliance.
D. Service level monitoring can only identify operational issues in the enterprise's operational environment.

Note: For more self-assessment questions, you may also want to obtain a copy of the *CRISC™ Review Questions, Answers & Explanations Manual 4th Edition* or the Database, which each consist of 500 multiple-choice study questions, answers and explanations.

SUGGESTED RESOURCES FOR FURTHER STUDY

In addition to the resources cited throughout this manual, the following resources are suggested for further study in this domain (publications in **bold** are stocked in the ISACA Bookstore):

Beasley, Mark S.; Bruce C. Branson; Bonnie V. Hancock; *Developing Key Risk Indicators to Strengthen Enterprise Risk Management*, Committee of Sponsoring Organizations of the Treadway Commission (COSO), USA, 2010

ISACA, *COBIT® 5 for Risk*, USA, 2013, *www.isaca.org/cobit*

ISACA, *COBIT® Self-assessment Guide: Using COBIT® 5*, USA, 2013, *www.isaca.org/cobit*

ISACA, *The Risk IT Practitioner Guide*, USA, 2009

Section Two: Content

4.0 OVERVIEW

The organization relies on its monitoring and reporting functions to identify risk for assessment and mitigation. The risk practitioner is able to best manage risk when monitoring is broad enough to provide a reasonable view of the risk environment but not so broad that the results are lost in a flood of data. Indicators for both performance and risk should be carefully considered and deliberately chosen based on their alignment with organizational goals. Identification and use of key risk indicators (KRIs) and key performance indicators (KPIs) can greatly improve the process of continuous monitoring. In addition, periodic assessments and testing may also be necessary as a means of identifying new and emerging risk. Because of the changing nature of risk and associated controls, ongoing monitoring is an essential step of the risk management life cycle as seen in **figure 4.1**.

Figure 4.1—The IT Risk Management Life Cycle

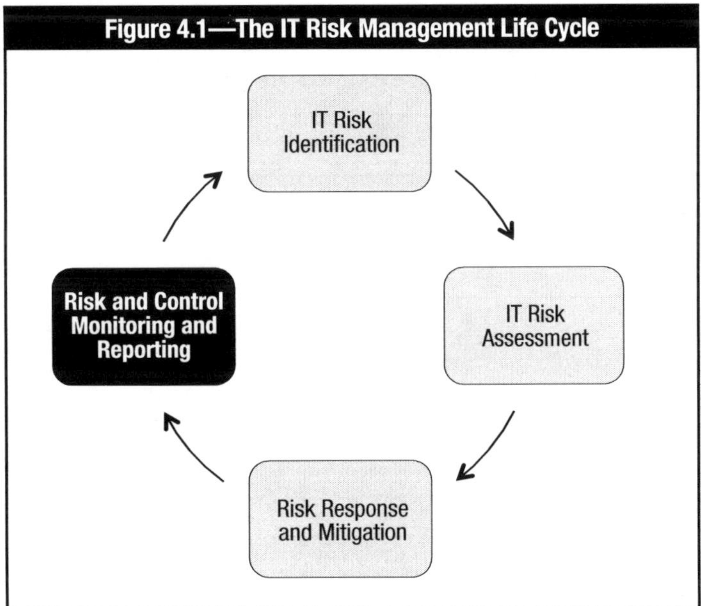

Monitoring is essential, but its effectiveness depends in large part on its successful integration with reporting. Consistent, repeatable methods of reporting provide management with a means of assessing the efficacy of the risk response and mitigation activities and justify the expenses of supporting security controls. Reports also assist management in exercising due care and due diligence in protecting the assets of the organization and meeting regulatory requirements.

Management, and the risk practitioner, should be aware of changes in risk over time and the impact of changes in the risk environment. As the goals and strategies of the organization evolve, it is natural for the risk environment to also evolve. Risk should be taken into account as part of strategic planning so that changes can be anticipated and mitigated as new strategic plans are unveiled, designed and implemented, rather than attempting to address these retroactively as part of the risk response and mitigation process.

4.1 KEY RISK INDICATORS

Risk indicators are used to measure risk levels in comparison to defined risk thresholds, so that the organization receives an alert when a risk level approaches an unacceptable level. By putting tracking and reporting mechanisms in place that alert staff to a developing or potential risk, the organization gains the opportunity to respond to the risk before it produces unacceptable outcomes.

KRIs comprise a subset of risk indicators that are highly relevant and possess a high probability of predicting or indicating important risk. Examples of KRIs include:
- Quantity of unauthorized equipment or software detected in scans
- Number of instances of service level agreements (SLAs) exceeding thresholds
- High average downtime due to operational incidents
- Average time to deploy new security patches to servers
- Excessive average time to research and remediate operations incidents
- Number of desktops/laptops that do not have current antivirus signatures or have not run a full scan within scheduled periods

KRIs support the following aspects of risk management:[1]
- Risk appetite, by validating the organization's risk appetite and risk tolerance levels
- Risk identification, by providing an objective means for identifying risk
- Risk mitigation, by providing a trigger for investigating an event or providing corrective action
- Risk culture, by helping the organization focus on important, relevant areas
- Risk measurement and reporting, by providing objective and quantitative risk information
- Regulatory compliance, by providing data that can be used as an input for operation risk capital calculations

4.1.1 KRI Selection

KRIs should be selected carefully and sparingly. Common mistakes made when implementing KRIs include regarding too many risk indicators as being KRIs and choosing KRIs that are flawed in some way. These include those KRIs that:
- Are not linked to specific risk
- Are incomplete or inaccurate due to unclear specifications
- Are difficult to measure, aggregate, compare and interpret
- Provide results that cannot be compared over time
- Are not linked to goals

The effectiveness of KRIs depends in large part on the strength of their metrics. Good metrics are SMART:
- **Specific**—Based on a clearly understood goal; clear and concise
- **Measureable**—Able to be measured; quantifiable (objective), not subjective
- **Attainable**—Realistic; based on important goals and values
- **Relevant**—Directly related to a specific activity or goal
- **Timely**—Grounded in a specific time frame

Other factors that can influence the selection of KRIs include:
- **Balance**—Risk indicators should be balanced and cover:
  - Lag indicators (indicating risk after events have occurred)
  - Lead indicators (indicating which controls are in place to prevent events from occurring)
  - Trends (analyzing indicators over time or correlating indicators to gain insights)
- **Root cause**—Selected indicators should drill down to the root cause of events, not just the symptoms.

The selection of an appropriate set of KRIs benefits the organization by:
- Providing an early warning (forward-looking) signal that a high risk is emerging to enable management to take proactive action (before the risk actually becomes a loss)
- Providing a backward-looking view on risk events that have occurred, enabling risk responses and management to be improved
- Enabling the documentation and analysis of trends
- Providing an indication of the enterprise's risk appetite and tolerance through metric setting (i.e., KRI thresholds)
- Increasing the likelihood of achieving the enterprise's strategic objectives
- Assisting in continually optimizing the risk governance and management environment

When working with management to determine appropriate KRIs, the risk practitioner should work with all relevant stakeholders to ensure greater buy-in and ownership. Risk indicators should be identified for all stakeholders, and IT-based metrics should be aligned with other metrics used in the organization to the greatest extent possible.

4.1.2 KRI Effectiveness

KRI effectiveness takes into consideration the following criteria:

- **Impact**—Indicators of risk with high business impact are more likely to be KRIs.
- **Effort**—For different indicators that are equivalent in sensitivity, the one that is easier to measure and maintain is preferred.
- **Reliability**—The indicator must possess a high correlation with the risk and be a good predictor or outcome measure.
- **Sensitivity**—The indicator must be representative of risk and capable of accurately indicating risk variances.
- **Repeatable**—A KRI must be repeatable and able to be measured on a regular basis to show trends and patterns in activity and results.

4.1.3 KRI Optimization

To ensure accurate and meaningful reporting, KRIs must be optimized to ensure that: (1) the correct data are being collected and reported on; and (2) the KRI thresholds are set correctly. KRIs that report on data points that either cannot be controlled by the enterprise or are not alerting management at the correct time to an adverse condition should be adjusted to be more precise, relevant or accurate.

Figure 4.2 describes examples of KRI optimization.

Figure 4.2—Examples in Which KRIs Should Be Optimized

Metric Criterion	Description
Sensitivity	Management has implemented an automated tool to analyze and report on access control logs based on severity, and the tool generates an excessive number of results. Management performs a risk assessment and decides to configure the monitoring tool to report only on alerts marked "critical."
Timing	Management has implemented strong segregation of duties (SoD) within the enterprise resource planning (ERP) system. One monitoring process tracks system transactions that violate the defined SoD rules before month-end processing is completed so that suspicious transactions can be investigated before reconciliation reports are generated.
Frequency	Management has implemented a key control that is performed multiple times a day. Based on a risk assessment, management decides that the monitoring activity can be performed weekly because this will capture a control failure in sufficient time for remediation.
Corrective action	Management has implemented a remediation process to bring controls into alignment with the organizational risk appetite. Using existing problem management tools, management is able to integrate automated monitoring of the controls in the process to prioritize existing gaps, assign problem owners and track remediation efforts.

4.1.4 KRI Maintenance

Because the organization's internal and external environments are constantly changing, the risk environment is also highly dynamic. The set of KRIs needs to be evaluated on a regular basis to verify that each KRI remains properly related to the risk appetite and tolerance levels of the enterprise, and that KRI trigger levels should be defined at a point that enables stakeholders to take appropriate action in a timely manner. Any KRIs that are no longer related to the risk appetite and tolerance should be replaced, while any whose trigger levels are found to be out of alignment with the requirements of the organization should be optimized.

4.2 KEY PERFORMANCE INDICATORS

Performance indicators measure how well a process is performing in terms of its stated goal. Like risk indicators, performance indicators are put in place in order to provide insight into whether action may be required.

A KPI is a performance indicator that is effective in predicting whether a goal will be reached and indicates the capabilities, practices and skills of value to the organization. KPIs measure activity goals, which are actions that process owners must take to achieve effective process performance. KPIs are used to set benchmarks for risk management goals and to monitor whether those goals are being attained. Management sets its risk management goals according to its risk acceptance level and desired cost-benefit analysis. For example, a KPI may indicate that an error rate of five percent is acceptable, which implies that an error rate higher than five percent is unacceptable and requires escalation with some form of response.

A KPI should be based on SMART metrics (see section 4.1 Key Risk Indicators) and should be:
- Valuable to the business
- Tied to a business function or service
- Under the control of management
- Quantitatively measured
- Used repeatedly in different reporting periods

Examples of potential KPIs include:
- Network availability
- Customer satisfaction
- Number of complaints resolved on first contact
- Time between data request and presentation
- Number of employees that attended awareness sessions

KPIs place emphasis on processes that should be good indicators of the health of the overall process. For example, the time required to deploy security patches is controllable by management, which can set policies, manage the change control process, mandate objectives and control resourcing. It is also quantitative and easily measured on a monthly or semiannual basis, so the results can be compared over time. KPIs are often used on charts or graphs to report compliance to management in a clear, easily understood manner.

4.2.1 Using KPIs with KRIs

KPIs and KRIs are often used in conjunction with one another to measure performance and mitigate risk. KPIs help to identify underperforming aspects of the organizations and areas of the business that may require additional resources and attention, while KRIs provide early warnings of increased risk within the organization.[2] Failing to meet performance goals is a risk, so the two can be used together in risk management or combined with indicators specific to controls (also called key control indicators [KCIs]) that may foreshadow an increase in risk due to a decrease in the effectiveness of mitigation.

To illustrate the difference, consider the risk that unpatched systems may lead to a serious breach of customer data, loss of availability and financial loss. In response to this risk, the organization develops a policy to apply all critical patches within 30 days. The development of this policy is an example of a KPI: the organization has a clear, measurable standard of what constitutes proper performance. On a scheduled basis, the organization tracks and reports to senior management on whether that KPI has been met and reports on the average time to deploy critical security patches.

If an organization tracks the time between the release of a patch and its deployment within the organization on a monthly basis, then a report that shows the average time taken could be developed, as shown in **figure 4.3**.

Figure 4.3—Example of Reporting KRIs and KPIs

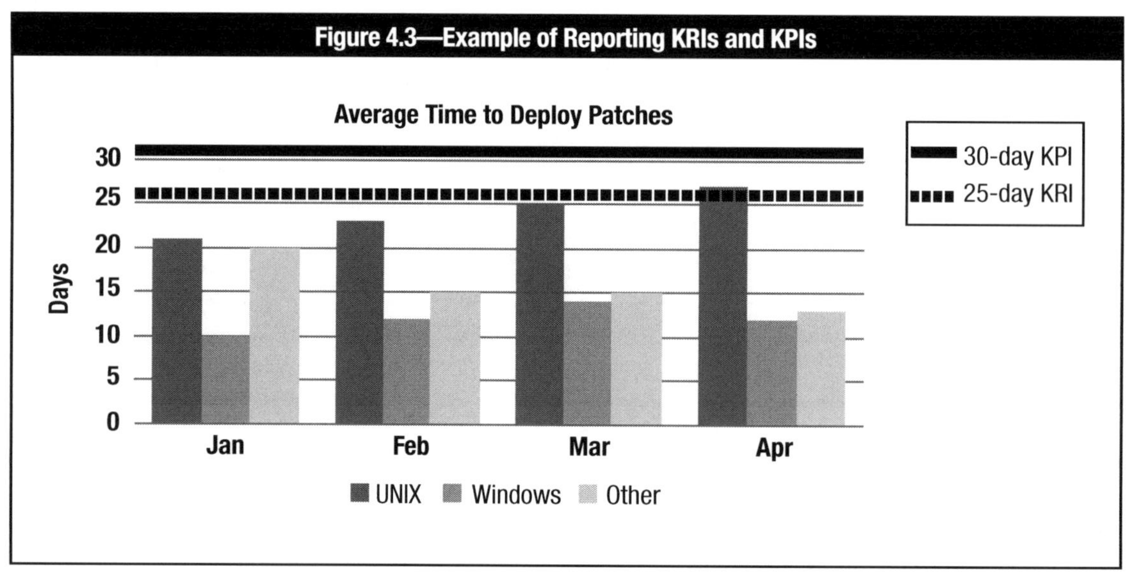

Through this chart, a manager can see that the time to deploy patches on a UNIX system is increasing each month and is in danger of exceeding the 30-day threshold. To detect a trend or developing problem, the organization may also set a KRI that will represent a threshold or level of performance that is getting dangerously close to exceeding the KPI but has not yet done so. By setting a KRI of 25 days, management can see that the time to deploy a UNIX patch has now increased to the point of triggering an alert. This alert should precipitate a review of the UNIX patching procedures, and management should investigate the situation to enable preventive action and address the slow deployment process before the deployments of these patches violates the policy.

This example shows how KRIs and KPIs can be used to monitor and track the attainment of goals. When measuring the success of a risk management program, KPIs must be based on meaningful criteria that can help management track the overall success of the controls put in place to support risk response.

4.3 DATA COLLECTION AND EXTRACTION TOOLS AND TECHNIQUES

The risk practitioner may use various sources of data to monitor and report on risk. Common internal data sources include:
- Prior risk assessments
- Project documents, especially risk logs and documented lessons learned
- Tickets from change, problem, release, configuration, asset and incident management systems
- Audit and incident reports
- User feedback and observation
- Interviews with management
- Security and test reports, including reports from tests of continuity and recovery plans
- Event and activity logs

4.3.1 Logs

Logging is commonly provided by systems, devices and applications and remains the most consistently popular way to capture and store data for analysis. Analysis of log data can identify security violations and be instrumental in forensics investigations. Log analysis can also alert the organization to malicious activity, such as a developing attack or multiple attempts to break in, and may be used to identify the source of an attack and assist in strengthening controls where necessary.

Logging has traditionally presented a trade-off between speed, detail and utility. If a log contains too many data from too many disparate sources, it may be difficult to notice significant individual events. Time synchronization of log entries can assist with correlation of events from multiple sources and improve the usefulness of review. Logging also takes time, potentially decreasing throughput for each transaction monitored. The risk practitioner should understand the ways in which log data is obtained in order to anticipate the extent to which the data captured presents a complete picture of activity.

Analysis of log data and control activity should answer the following questions:
- Are the controls operating correctly?
- Is the level of risk acceptable?
- Are the risk strategy and controls aligned with business strategy and priorities?
- Are the controls flexible enough to meet changing threats?
- Is the correct risk data being provided in a timely manner?
- Is the risk management effort aiding in reaching corporate objectives?
- Is the awareness of risk and compliance embedded into user behaviors?

As network and client security systems continue to mature, logging for specific purposes is becoming more granular and offering a greater degree of detail for analysis. For instance, client-resident data leakage protection (DLP) software can integrate with antivirus and anti-malware modules to distinguish between interactive (human) and process-driven attempts to migrate data in ways that are forbidden. Such information might reveal the difference between inadequate user training (or a malicious insider) and a remotely compromised system. Logging by intrusion detection or prevention systems (IDS/IPS) placed within the network can also be useful in detecting suspicious traffic patterns, particularly when combined with advanced behavior-based (heuristic) analysis.

Logs may contain information that is sensitive or needed for forensic purposes, so they should be configured in ways that prevent alteration or deletion as well as preventing access by authorized personnel. In particular, administrators with responsibility for systems or applications should generally not have the ability to alter or delete logs made against their own scopes of responsibility. The risk practitioner should consider log access permissions as part of evaluating the potential insider threat.

4.3.2 Security Information and Event Management

As the numbers of network end points and routes expand, the risk practitioner may become overwhelmed by the sheer volume of data presented by logs and other individually based alerting mechanisms. Security information and event management (SIEM) systems are integrated data correlation tools that address this problem by capturing data from multiple sources and analyzing the system, application and network activity reported in these data feeds for possible security events. SIEM systems can be used to detect attacks in progress by signature or behavior (heuristics) as well as identify compliance violations. **Figure 4.4** illustrates a notional SIEM architecture.

Figure 4.4—A SIEM System

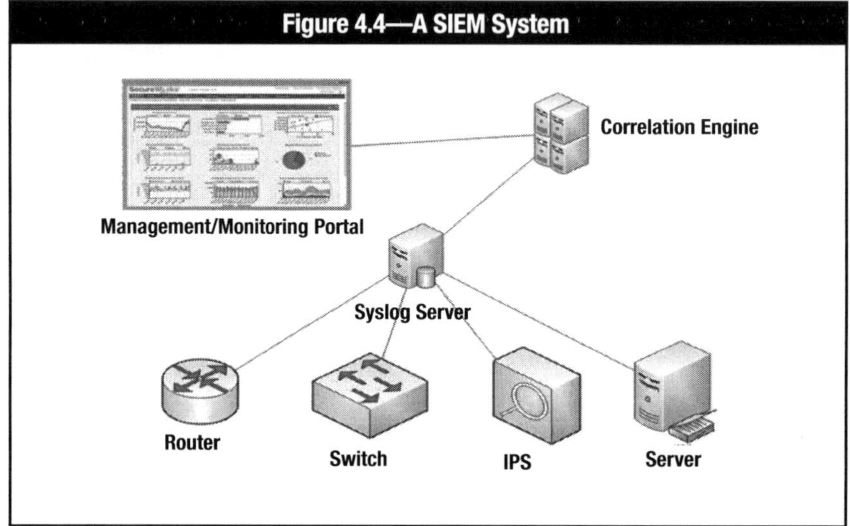

The key to SIEM is the speed of the correlation engine. By gathering and correlating data from multiple sources, a SIEM system can develop reports on security at the enterprise level while retaining the ability to highlight relationships among activity on various parts of the network or systems. Correlation based on multiple criteria such as type, timing, chronological sequence and identified source of an event are all commonly available in SIEM implementations. In many cases, very granular assessment is possible including activities associated with specific transaction codes, network IDs or types of traffic that may be generally permissible but are outside of common behavioral norms for the monitored network.

By highlighting developing trends, SIEM systems allow risk practitioners to identify risk and bring it to the attention of management before it would typically be discernable at the level of individual logs.

4.3.3 Integrated Test Facilities

When seeking to monitor the performance and operation of an application, one common method is to use an integrated test facility (ITF). An ITF is a testing methodology that processes test data through production systems to test whether the systems are operating correctly. Within an ITF, the organization might set up several fictitious customers or transactions that are processed along with real data, allowing business analysts to observe the operation of the production systems and ensure correct processing.

4.3.4 External Sources of Information

Although internal sources provide the clearest picture of what is happening within the organization, the risk practitioner can gain additional insight by reviewing external data sources such as:

- Media reports
- Computer emergency (or incident) response team (CERT/CIRT) advisories
- Security company reports
- Regulatory bodies
- Peer organizations

Reports from antivirus and security companies, or from telecommunications providers such as the Verizon Data Breach Investigations report, are issued annually to provide a review of risk and exposure factors and may assist in developing and presenting a risk management solution to management. Information on incidents and risk may also come from computer or cybersecurity monitoring and reporting services by government and nonprofit organizations.

4.4 MONITORING CONTROLS

The risk practitioner should work with stakeholders to set up an IS control monitoring process that reflects objectives, scopes and methods consistent with enterprise objectives and integrate this approach with performance management at the enterprise level.[3]

In order to prevent an acceptance of risk within IT that exceeds the risk acceptance criteria set by the business, risk monitoring and evaluation should be integrated with enterprise performance management systems to ensure alignment between IT risk and business risk. The controls mandated through risk management must be aligned with IT security and related enterprise policies, subject to a regular process of review and revision to ensure that IT policies support the requirements of the organization.

The purpose of the IS control monitoring function is to ensure that IT security requirements are being met; standards are being followed; and staff is complying with the policies, practices and procedures of the organization.

The monitoring and metrics of managing a control must be based on data relevant to the risk and the overall performance of the device.

Monitoring controls is a process that has six steps:

1. Identify and confirm risk control owners and stakeholders.
2. Engage with stakeholders and communicate the risk and information security requirements and objectives for monitoring and reporting.
3. Align and continually maintain the information security monitoring and evaluation approach with the IT and enterprise approaches.
4. Establish the information security monitoring process and procedure.
5. Agree on a life cycle management and change control process for information security monitoring and reporting.
6. Request, prioritize and allocate resources for monitoring information security.

The risk practitioner should remember that because the purpose of a control is to mitigate a risk, the purpose of control monitoring is to verify whether the control is effectively addressing the risk, not to see whether the control itself is working. For example, a firewall that is configured in a way that does not regulate traffic between two networks may be ineffective even if the firewall is operational and processing traffic.

Risk monitoring and evaluation are processes performed to:

- Collect, validate and evaluate business, IT and process goals and metrics
- Monitor processes to ensure that they are performing in line with established performance metrics
- Provide reports that are systematic and timely

To accomplish these tasks, data related to risk management must be gathered from various sources in a timely and accurate manner. After the data have been validated for integrity, analysis can be performed against specific performance targets. Properly done, this process provides a succinct, all-around view of IT performance within the enterprise monitoring system.

The risk practitioner should continuously monitor, benchmark and improve the IT control environment and control framework to meet organizational objectives. When the results of the monitoring indicate an area of noncompliance or unacceptable performance, the risk practitioner should recommend mitigation activities such as implementation of new controls, adjustment or enforcement of existing controls, or changes to a business process to address the risk environment.

The monitoring of controls and the risk management framework may be done through self-assessment or independent assurance reviews. When risk action plans are required to address a risk, the plans should be monitored to ensure appropriate risk management practices in alignment with risk appetite and tolerance. Assurance activities performed by independent entities always should be independent from the function, group or organization being monitored, and these entities should have the necessary skills and competence to perform assurance and adhere to codes of ethics and professional standards.

Any exceptions to risk monitoring and control activities should be reported promptly, followed up with analysis and addressed with properly prioritized corrective actions. As the risk environment changes and internal control systems are affected by changes in business and IT risk, gaps in the risk environment must be evaluated and changes made where necessary.

Sources of control monitoring information include:
- Network operations centers (NOC)
- Command and control centers
- Continuous control monitoring
- Periodic testing
- Independent assessments

The risk practitioner should encourage each process owner to take positive ownership of control improvement through a continuing program of self-assessment to evaluate the completeness and effectiveness of management's control over processes, policies and contracts. In addition, explicit acknowledgement by management that all monitoring activities are functioning as designed should be recorded in the risk registry.

4.5 CONTROL ASSESSMENT TYPES

Effective control monitoring depends on the accuracy and completeness of the data provided for evaluation. The risk practitioner must ensure that the data received are genuine and free from errors or misstatements. Data that can be retrieved directly by the risk practitioner are preferable to data provided by a third party. The risk practitioner should encourage local ownership of risk and control monitoring, which helps instill a risk culture in which local managers accept responsibility for risk and enables a faster detection of violations and security incidents.

4.5.1 IS Audit

Audit teams provide independent and objective review of the effectiveness and appropriateness of the control environment. Information provided by IS auditors can underline the need for control enhancement and bring risk to the attention of management. By working with audit teams, the risk practitioner can align the risk management program with the audit program and may be able to provide supporting data to the IS auditors. Recommendations provided by IS audit in turn will often require the attention of the risk practitioner through the updating of risk action plans and the risk register as well as enhancement of controls.

4.5.2 Vulnerability Assessment

A vulnerability assessment is a methodical review of security to ensure that there are no predictable and unaddressed attack vectors (such as unnecessary open ports or services) that could be used to intentionally or unintentionally compromise an environment. The scope of an assessment may range from a single system or application to an entire facility or end-to-end business process. The purpose of a vulnerability assessment is to deliver information to management that can be used both to understand the effectiveness of the risk management program and make decisions regarding treatment of identified vulnerability, such as deploying new controls.

A risk practitioner participating in a vulnerability assessment should have a working knowledge of the existing security controls, both technical and nontechnical, as well as the overall architecture. For example, a review of a firewall may include the configuration (technical control), the change control process for managing the configuration (nontechnical control), the process for monitoring the firewall logs, the system architecture to ensure the firewall cannot be bypassed, and the training and competence levels of the firewall administrators. There are many open source and commercial tools that can be used to automate or supplement certain aspects of the vulnerability assessment process. Additionally, many web sites list known vulnerabilities with common applications, operating systems (OSs) and utilities.

Regular, rigorous vulnerability assessment is an important tool in the risk management process, but assessment is not unto itself a comprehensive solution. Vulnerabilities identified by an assessment may be false positives or may be addressed by compensating controls, and even genuine vulnerabilities may be accepted by management if the only available mitigation strategies cost more than the potential impact of compromise. Also, even the most rigorous vulnerability assessment is limited to weaknesses that are either generally known or can be discerned by the assessor based on the information at hand. Risk practitioners should be careful to avoid the trap of reliance upon automated tools, particularly when assessing vulnerability of IT systems. New methods of compromising systems are continuously being discovered, both intentionally and unintentionally, and scanning detects only those weaknesses whose use in compromising other systems has already been publicized. Automated tools are best viewed as a way to save time on certain areas of assessment so the practitioner can devote more attention to other areas.

4.5.3 Penetration Testing

A penetration test is a targeted attempt to break into an environment. As with a vulnerability assessment, the target of a penetration test may be a system, application, facility or end-to-end business process.

If the goal of a penetration test is to validate a vulnerability assessment, the tester is typically given the results of the assessment in order to focus on attempts to exploit the vulnerabilities that have already been identified. If the penetration tester is able to break in, then the vulnerability is real and must be mitigated; otherwise, there is a good chance that the vulnerability may be a false positive and not require mitigation. This validation-centric approach is typically called a white hat penetration test.

Penetration tests can also be undertaken against systems believed to be secure. In order to test the effectiveness of the security associated with such a system, the tester is generally given no information beyond a limited number of ground rules for the engagement and a definition of what constitutes a successful compromise of the system. This approach closely approximates the circumstances of someone attempting to break into a system maliciously and is typically called a black hat penetration test.

In carrying out his or her work, a penetration tester often uses the same tools used by malicious hackers to try to break into systems. This approach provides meaningful results, but methods used for penetration testing may pose a risk to the organization in the form of system failure or data compromise. Therefore, it is of utmost importance that such tests—whether white hat or black hat—are conducted only with management approval, using a defined methodology and under proper oversight.

The results of the penetration test should be provided to management for follow up and review.

4.5.4 Third-party Assurance

An organization can earn customer and shareholder confidence by having a third party provide assurance or attestation of the validity of the information security program of the. Third-party attestation may be based on an external IS audit or a certification of compliance with an internationally recognized standard, such as COBIT 5 or *ISO/IEC 27001—Information security management*, or an industry standard, such as the Payment Card Industry Data Security Standard (PCI DSS). The third party is responsible for evaluating the processes of the subject organization and validating compliance with the requirements of the standard.

Another source for attestation is based on the Statement on Standards for Attestation Engagements Number 16 (SSAE 16). SSAE 16 was developed by the American Institute of Certified Public Accountants (AICPA) to replace the older SAS 70 standard. SSAE 16 complies with the international service organization reporting standard International Standards for Assurance Engagements Number 3402 (ISAE 3402). This form of third-party assurance is used by many organizations when relying on cloud or third-party service suppliers, where regular or continuous monitoring is particularly important to establish stakeholder confidence in the services being delivered outside the control of the organization.

4.6 RESULTS OF CONTROL ASSESSMENTS

The risk practitioner is expected to provide a report to management on the status of the risk management program and the overall risk profile of the organization. Making such a report requires the review of the effectiveness of the controls in the organization and their compliance with established policy. Controls may need adjustment, replacement or removal depending on the changes in the risk environment and the acceptance and appropriateness of the controls.

The effectiveness of control monitoring is dependent on the following:
- Timeliness of the reporting—Are data received in time to take corrective action?
- Skill of the data analyst—Does the analyst have the skills to properly evaluate the controls?
- Quality of monitoring data available—Are the monitoring data accurate and complete?
- Quantity of data to be analyzed—Can the risk practitioner find the important data in the midst of all the other log data available?

4.6.1 Maturity Model Assessment and Improvement Techniques

The risk practitioner should be committed to continuous improvement of the risk management program. A mature and healthy risk management program will be better at preventing, detecting and responding to security events and risk scenarios, and maturity and growth comes from practice and attention to learning from past events. As the risk practitioner develops the skills, tools and team necessary for better risk management, the consistency in how risk is identified, assessed, mitigated and monitored will also mature and develop.

Routine use of a capability maturity model (CMM) shows the maturation of the risk management process year over year. A CMM starts with level zero—undefined and *ad hoc* activities and progresses—through the steps of defining and following a program; learning and enhancement of the program; and finally, a mature program that represents stable, quality processes and reliable, accurate information.

An example of the CMM is shown in **figure 4.5**.

Figure 4.5—Example of the Capability Maturity Model

COBIT 5 ISO/IEC 15504-based Capability Levels	Meaning of the COBIT 5 ISO/IEC 15504-based Capability Levels	Context
5 Optimized	The previously described predictable process is continuously improved to meet relevant current and projected business goals.	Enterprise view/Corporate knowledge
4 Predictable	The previously described established process now operates within defined limits to achieve its process outcomes.	
3 Established	The previously described managed process is now implemented using a defined process that is capable of achieving its process outcomes.	
2 Managed	The previously described performed process is now implemented in a managed fashion (planned, monitored and adjusted) and its work products are appropriately established, controlled and maintained.	Instance view/Individual knowledge
1 Performed	The implemented process achieves its process purpose.	
0 Incomplete	The process is not implemented or fails to achieve its process purpose. At this level, there is little or no evidence of any systematic achievement of the process purpose.	

Adapted from ISACA, COBIT 5, USA, 2012, figure 20

4.7 CHANGES TO THE IT RISK PROFILE

Risk management serves the needs of the organization. Because the organization evolves over time, risk management is an ongoing, cyclical process that recognizes the dynamic nature of risk and the need for continuous monitoring and assessment. Temporary stability may be attained as a result of program maturity, but the risk practitioner should not become complacent. Changes in the organizational risk environment may prompt complete reevaluation at any time.

The risk profile is based on the overall risk posture of the organization reflected in its attentiveness to monitoring the effectiveness of controls, proactivity in identifying and addressing or preventing risk and development of a risk culture.

Changes in the risk profile may be the result of numerous factors, including:
- New technologies
- Changes to business procedures
- Mergers or acquisitions
- New or revised regulations
- Changes in customer expectations
- Actions of competitors
- Effectiveness of risk awareness programs

Management should pursue the development and maturity of risk management processes and culture with careful attention. Risk at the organization level is the product of risk as it exists at the levels of individual systems, facilities and process. Similar to health, safety and information security, the risk posture of an organization may be affected by the cascading effects of minor actions.

To evaluate the risk control effectiveness and efficiency and determine the IT risk profile of the organization, risk practitioners should continually monitor an assortment of indicators including:
- New assets
- Changes to the scope of risk assessment
- Changes in business priorities, assets, services, products and operations
- Risk acceptance levels
- New threats (internal and external)
- Newly discovered vulnerabilities
- Possibility of increased risk due to aggregation of threats and vulnerabilities
- Incidents
- Logs and other data sources
- People and the morale of the organization
- Regulations and legal changes

Other areas that may be candidates for monitoring include:
- Actions of competitors (benchmarking and standards)
- Changes in the supply chain (e.g., mergers, political turmoil, transportation problems, natural events, bankruptcies of vendors)
- Changes in the financial markets
- Total cost of ownership (TCO) of assets
- Impact from external events
- Availability of staff/resources

IT risk management objectives and goals should be reviewed to ensure that they continue to be aligned with the goals and objectives of senior management on a regular basis. Many organizations find it beneficial to perform the review on an annual basis and to include criteria for monitoring, thresholds used for KPIs and KRIs, policies and strategies of risk, the reporting schedule, and the list of key stakeholders to be notified when KPIs or KRIs exceed their thresholds. The review also provides a valuable opportunity to review the program in terms of increasing maturity, including the completion of risk response and mitigation activities, the training of staff, the success of awareness programs, improved response time to incidents, timely rollout of patches, and better alignment and communication among management, audit, business continuity, physical security and information security departments.

Risk is owned by management, but the risk practitioner has a key role in ensuring that management is aware of the current IT risk profile of the organization and that risk is being managed in a way that meets management objectives. Throughout this phase of the IT risk management process, the risk practitioner works with risk owners, IT staff, third parties, incident response teams and auditors to monitor risk and evaluate the effectiveness and efficiency of the control framework. As incidents occur, lessons learned are used to improve the risk management process through better knowledge, staffing, technical controls, procedures, monitoring and response programs. These benefits can help avoid future problems, enable the impact of future incidents to be minimized and sustain business operations.

4.8 SUMMARY

Proper and effective management of risk is essential to protecting the assets and attaining the goals of the organization. Throughout this review manual, the principles of risk management have been discussed along with the practical steps used to demonstrate the benefits, objectives and concepts associated with the identification, assessment, reporting, response and monitoring of risk. The objective of risk management is to discover and address all risk in an appropriate manner and ensure that the organization has reduced risk to acceptable levels and does not become a victim of a risk that should have been identified and mitigated.

As risk is identified and assessed, risk owners select appropriate responses and create risk action plans to implement or modify controls selected to mitigate risk. The risk owner also mandates the development of the ability to monitor and report on the effectiveness of these controls, because without the ability to monitor the controls, there can be no assurance that the controls are working correctly or effectively mitigating the risk.

Regular monitoring and reporting on risk is essential to management, and the use of KPIs and KRIs assists management in the monitoring of trends, compliance and issues related to risk. Continuous monitoring of risk by every manager and every department, coupled with a robust risk culture that emphasizes local ownership and accountability, will aid in timely reporting and response to risk events and facilitate the development of more effective and efficient risk management.

Risk management is a never-ending process. As the external and internal environments change, as technology progresses, and the nature of attacks and attackers evolves, the need to revisit the risk management effort and reassess risk, revise risk response and improve the risk culture and awareness of risk throughout the organization is renewed. IT risk and controls should be continuously monitored and reported on to relevant stakeholders to ensure the continued efficiency and effectiveness of the IT risk management strategy and its alignment to business objectives.

ENDNOTES

[1] Fraser, John; Betty Simkins; *Enterprise Risk Management: Today's Leading Research & Best Practices for Tomorrow's Executives*, John Wiley & Sons, USA, 2010

[2] Beasley, Mark S.; Bruce C. Branson; Bonnie V. Hancock; *Developing Key Risk Indicators to Strengthen Enterprise Risk Management*, Committee of Sponsoring Organizations of the Treadway Commission (COSO), USA, 2010.

[3] ISACA, *COBIT 5 for Risk*, USA, 2013

GENERAL INFORMATION

The CRISC certification is designed to meet the growing demand for professionals who can integrate enterprise risk management (ERM) with discrete IS control skills. The technical skills and practices the CRISC certification promotes and evaluates are the building blocks of success in this growing field, and the CRISC designation demonstrates proficiency in this role.

Note: As information regarding the CRISC examination, requirements and locations and dates may change, please refer to *www.isaca.org/certification* for the most up-to-date information.

REQUIREMENTS FOR CERTIFICATION

To earn the CRISC designation, the following requirements must be met:

1. Pass the CRISC exam.
2. Submit an application (within five years of the passing date) with verified evidence of a minimum of at least three years of cumulative work experience performing the tasks of a CRISC professional across at least two CRISC domains. Of the two required domains, one must be risk-related, either Domain 1 (IT Risk Identification) or 2 (IT Risk Assessment). There will be no substitutions or experience waivers. A processing fee must accompany all applications.
3. Adhere to the ISACA Code of Professional Ethics.
4. Agree to comply with the CRISC continuing education policy.

Please note that certification application decisions are not final as there is an appeal process for certification application denials. Appeals undertaken by a certification exam taker, certification applicant or by a certified individual are undertaken at the discretion and cost of the exam taker, applicant or individual. Inquiries regarding denials of certification can be sent to *certification@isaca.org.*

SUCCESSFUL COMPLETION OF THE CRISC EXAM

The exam is open to all individuals who wish to take it. Successful exam candidates are not certified until they apply for certification (and demonstrate that they have met all requirements) and receive approval from ISACA.

EXPERIENCE IN RISK AND INFORMATION SYSTEMS CONTROL

Work experience must be gained within the 10-year period preceding the application for certification or within five years from the date of initially passing the exam. An application for certification must be submitted within five years from the passing date of the CRISC exam. All experience must be verified independently with employers.

Note: A CRISC candidate may choose to take the CRISC exam prior to meeting the experience requirements.

DESCRIPTION OF THE EXAM

The CRISC Certification Working Group oversees the development of the exam and ensures the currency of its content. The exam consists of 150 multiple-choice questions that cover the CRISC job practice domains. The job practice was developed and validated using prominent industry leaders, subject matter experts and industry practitioners.

REGISTRATION FOR THE CRISC EXAM

The CRISC exam will be administered twice annually. Please refer to the *ISACA Exam Candidate Information Guide* at *www.isaca.org/examguide* for specific registration dates and deadlines, as well as important key information for exam day. Exam registrations can be placed online at *www.isaca.org/examreg.*

CRISC PROGRAM ACCREDITATION UNDER ISO/IEC 17024:2012

The American National Standards Institute (ANSI) has voted to continue the accreditation for the CISA, CISM, CGEIT and CRISC certifications under ISO/IEC 17024:2012, General Requirements for Bodies Operating Certification Systems of Persons. ANSI, a private, nonprofit organization, accredits other organizations to serve as third-party product, system and personnel certifiers.

ISO/IEC 17024 specifies the requirements to be followed by organizations certifying individuals against specific requirements. ANSI describes ISO/IEC 17024 as "expected to play a prominent role in facilitating global standardization of the certification community, increasing mobility among countries, enhancing public safety, and protecting consumers."

ANSI's accreditation:
- Promotes the unique qualifications and expertise ISACA's certifications provide
- Protects the integrity of the certifications and provides legal defensibility
- Enhances consumer and public confidence in the certifications and the people who hold them
- Facilitates mobility across borders or industries

Accreditation by ANSI signifies that ISACA's procedures meet ANSI's essential requirements for openness, balance, consensus and due process. With this accreditation, ISACA anticipates that significant opportunities for CISAs, CISMs, CGEITs and CRISCs will continue to open in the United States and around the world.

PREPARING FOR THE CRISC EXAM

The CRISC exam evaluates a candidate's practical knowledge of the job practice domains listed in this manual and online at *www.isaca.org/criscjobpractice*. That is, the exam is designed to test a candidate's knowledge and experience of the proper application of IT risk and IS control best practices. Since the exam covers a broad spectrum of risk and IS control issues, candidates are cautioned not to assume that reading CRISC study guides and reference publications will fully prepare them for the exam. CRISC candidates are encouraged to refer to their own experiences when studying for the exam and refer to CRISC study guides and reference publications for further explanation of concepts or practices with which the candidate is not familiar.

No representation or warranties are made by ISACA in regard to CRISC exam study guides, other ISACA publications, references or courses assuring candidates' passage of the exam.

TYPES OF EXAM QUESTIONS

CRISC exam questions are developed with the intent of measuring and testing practical knowledge and the application of general concepts and standards. All questions are multiple choice and are designed for one best answer.

Every question has a stem (question) and four options (answer choices). The candidate is asked to choose the correct or best answer from the options. The stem may be in the form of a question or incomplete statement. In some instances, a scenario may also be included. These questions normally include a description of a situation and require the candidate to answer two or more questions based on the information provided. The candidate is cautioned to read each question carefully. An exam question may require the candidate to choose the appropriate answer based on a qualifier, such as **MOST** important or **BEST**. In every case, the candidate is required to read the question carefully, eliminate known incorrect answers and then make the best choice possible. To gain a better understanding of the types of question that might appear on the exam and how these questions are developed, refer to the Item Writing Guide available at *www.isaca.org/itemwriting.*

ADMINISTRATION OF THE EXAM

ISACA has contracted with an internationally recognized testing agency. This not-for-profit corporation engages in the development and administration of credentialing exams for certification and licensing purposes. It assists ISACA in the construction, administration and scoring of the CRISC exam.

SITTING FOR THE EXAM

Candidates are to report to the testing site at the time indicated on their admission ticket. NO CANDIDATE WILL BE ADMITTED TO THE TEST CENTER ONCE THE CHIEF EXAMINER BEGINS READING THE ORAL INSTRUCTIONS. Candidates who do not attend the scheduled exam date or arrive after the oral instructions have begun will not be allowed to sit for the exam and will forfeit their registration fee. To ensure that candidates arrive in time for the exam, it is recommended that candidates become familiar with the exact location of, and the best travel route to, the exam site prior to the date of the exam. Candidates can use their admission tickets only at the designated test center on the admission ticket.

To be admitted into the test site, candidates must bring the email printout OR a printout of the downloaded admission ticket and an acceptable form of photo identification such as a driver's license, passport or government ID. This ID must be a current and original government-issued identification that is not handwritten and that contains both the candidate's name as it appears on the admission ticket and the candidate's photograph. Candidates who do not provide an acceptable form of identification will not be allowed to sit for the exam and will forfeit their registration fee. Candidates are not to write on the admission ticket.

The following conventions should be observed when completing the exam:

- Do not bring study materials (including notes, paper, books or study guides) or scratch paper or notepads into the exam site. For further details regarding what personal belongings can (and cannot) be brought into the test site, please visit *www.isaca.org/criscbelongings.*
- Candidates are not allowed to bring any type of communication, surveillance or recording device (including, but not limited to, cell phones, tablets, smart glasses, smart watches, mobile devices, etc.) into the test center. If candidates are viewed with any such device during the exam administration, their exams will be voided and they will be asked to immediately leave the exam site.
- Candidates who leave the testing area without authorization or accompaniment by a test proctor will not be allowed to return to the testing room and will be subject to disqualification.
- Candidates should bring several no. 2 pencils since pencils will not be provided at the exam site.
- As exam venues vary, every attempt will be made to make the climate control comfortable at each exam venue. Candidates may want to dress to their own comfort level.
- Read the provided instructions carefully before attempting to answer questions. Skipping over these directions or reading them too quickly could result in missing important information and possibly losing credit points.
- Mark the appropriate area when indicating responses on the answer sheet. When correcting a previously answered question, fully erase a wrong answer before writing in the new one.
- Remember to answer all questions since there is no penalty for wrong answers. Grading is based solely on the number of questions answered correctly. Do not leave any question blank.
- Identify key words or phrases in the question (**MOST**, **BEST**, **FIRST** …) before selecting and recording the answer.
- The chief examiner or designate at each test center will read aloud the instructions for entering information on the answer sheet. It is imperative that candidates include their exam identification number as it appears on their admission ticket and any other requested information on their exam answer sheet. Failure to do so may result in a delay or errors.

BUDGETING TIME

The following are time-management tips for the exam:

- It is recommended that candidates become familiar with the exact location of, and the best travel route to, the exam site prior to the date of the exam.
- Candidates should arrive at the exam testing site at the time indicated on their admission ticket. This will give the candidate time to be seated and get acclimated.

- The exam is administered over a four-hour period. This allows for a little over 1.5 minutes per question. Therefore, it is advisable that candidates pace themselves to complete the entire exam. In order to do so, candidates should complete an average of 37.5 questions per hour.
- Candidates are urged to record their answers on their answer sheet. No additional time will be allowed after the exam time has elapsed to transfer or record answers should candidates mark their answers in the question booklet. The exam will be scored based on the answer sheet recording only.

RULES AND PROCEDURES

- Candidates are asked to sign the answer sheet to protect the security of the exam and maintain the validity of the scores.
- Candidates who are discovered engaging in any kind of misconduct—including, but not limited to, giving or receiving help; using notes, papers or other aids; attempting to take the exam for someone else; using any type of communication, surveillance or recording device, including cell phones, during the exam administration; removing test materials, answer sheet or notes from the testing room; or attempting to share test questions or answers or other information contained in the exam (as such are the confidential information of ISACA)—will have their exam voided and be asked to leave the exam site. Candidates who leave the testing area without authorization or accompaniment by a test proctor will not be allowed to return to the testing room and will be subject to disqualification. Candidates who continue to write the exam after the proctor signals the end of the examination time may have their examination voided. Candidates may not access items stored in the personal belongings area until they have completed their exams. The testing agency will report all cases of misconduct to the respective ISACA Certification Working Group for review in order to render any decision necessary. Sharing the confidential test items subsequent to the exam will also be considered misconduct resulting in a voided examinations score.
- Candidates may not take the exam question booklet after completion of the exam.
- Candidates are not permitted to access items stored in the personal belongings area during the exam.
- The *ISACA Exam Candidate Information Guide* includes candidate information about exam registration, dates and deadlines and provides important key candidate details for exam day administration. This publication is available online at *www.isaca.org/examguide*. We encourage exam candidates to review the information in this guide to familiarize themselves with the rules for exam day.

GRADING THE CRISC EXAM AND RECEIVING RESULTS

The exam consists of 150 items. Candidate scores are reported as a scaled score. A scaled score is a conversion of a candidate's raw score on an exam to a common scale. ISACA uses and reports scores on a common scale from 200 to 800. A candidate must receive a score of 450 or higher to pass the exam. A score of 450 represents a minimum consistent standard of knowledge as established by ISACA's CRISC Certification Working Group. A candidate receiving a passing score may then apply for certification if all other requirements are met.

Passing the exam does not grant the CRISC designation. To become a CRISC, each candidate must complete all requirements, including submitting an application for certification.

The CRISC examination contains some questions which are included for research and analysis purposes only. These questions are not separately identified and the candidate's final score will be based only on the common scored questions.

A candidate receiving a score less than 450 is not successful and can retake the exam by registering and paying the appropriate exam fee for any future exam administration. To assist with future study, the result letter each candidate receives includes a score analysis by content area. There are no limits to the number of times a candidate can take the exam.

Approximately eight weeks after the test date, the official exam results will be mailed to candidates. Additionally, with the candidate's consent during the registration process, an email containing the candidates pass/fail status and score will be sent to paid candidates. This email notification will only be sent to the address listed in the candidate's profile at the time of the initial release of the results. To ensure the confidentiality of scores, exam results will not be reported by telephone or fax. To prevent the email notification from being sent to the candidate's spam folder, the candidate should add *exam@isaca.org* to his/her address book, whitelist or safe senders list.

In order to become CRISC-certified, candidates must pass the exam and must complete and submit an application for certification (and must receive confirmation from ISACA that the application is approved) within five (5) years of the passing date. We encourage applicants to apply as soon as they have met the requirements. The application is available on the ISACA web site at *www.isaca.org/criscapp*. Please select the CRISC application that corresponds to your passing date. Once the application is approved, the applicant will be sent confirmation of the approval. The candidate is not CRISC-certified, and cannot use the CRISC designation, until the candidate's application is approved. A processing fee must accompany your CRISC application for certification.

The score report contains a subscore for each job practice domain. The subscores can be useful in identifying those areas in which the candidate may need further study before retaking the exam. Unsuccessful candidates should note that taking either a simple or weighted average of the subscores does not derive the total scaled score. Candidates receiving a failing score on the exam may request a rescoring of their answer sheet. This procedure ensures that no stray marks, multiple responses or other conditions interfered with computer scoring. Candidates should understand, however, that all scores are subjected to several quality control checks before they are reported; therefore, rescores most likely will not result in a score change. Requests for hand scoring must be made in writing to the certification department within 90 days following the release of the exam results. Requests for a hand score after the deadline date will not be processed. All requests must include a candidate's name, exam identification number and mailing address. A fee of US $75 must accompany this request.

CONFIDENTIALITY

By taking an ISACA Exam, the candidate understands and agrees that the Exam (which includes all aspects of the exam, including, without limitation, the test questions, answers and examples and other information presented or contained in the exam) belongs to ISACA and constitutes ISACA's confidential information (collectively, Confidential Information). The candidate agrees to maintain the confidentiality of all of ISACA 's Confidential Information at all times and understands that any failure to maintain the confidentiality of ISACA's Confidential Information may result in disciplinary action against the candidate by ISACA or other adverse consequences, including, without limitation, nullification of the candidate's exam, loss of the candidate's credentials and/or litigation. Specifically the candidate understands that he/she may not, for example, discuss, publish or share any exam question(s), the candidate's answer to any question(s) or the exam's format with anyone in any forum or media (i.e., via email, Facebook, LinkedIn or any other form of social media).

Page intentionally left blank

GLOSSARY

Note: Glossary terms are provided for reference within the CRISC Review Manual. As definitions of terms may evolve due to the changing technological environment, please see *www.isaca.org/glossary* for the most up-to-date terms and definitions.

A

Access control—The processes, rules and deployment mechanisms that control access to information systems, resources and physical access to premises

Access rights—The permission or privileges granted to users, programs or workstations to create, change, delete or view data and files within a system, as defined by rules established by data owners and the information security policy

Accountability—The ability to map a given activity or event back to the responsible party

Advanced persistent threat (APT)—An adversary that possesses sophisticated levels of expertise and significant resources which allow it to create opportunities to achieve its objectives using multiple attack vectors (NIST SP800-61)

Scope notes: The APT:
1. Pursues its objectives repeatedly over an extended period of time
2. Adapts to defenders' efforts to resist it
3. Is determined to maintain the level of interaction needed to execute its objectives

Application controls—The policies, procedures and activities designed to provide reasonable assurance that objectives relevant to a given automated solution (application) are achieved

Architecture—Description of the fundamental underlying design of the components of the business system, or of one element of the business system (e.g., technology), the relationships among them, and the manner in which they support enterprise objectives

Asset—Something of either tangible or intangible value worth protecting, including people, information, infrastructure, finances and reputation

Asset value—The value of an asset is subject many factors including the value to both the business and to competitors. An asset may be valued according to what another person would pay for it, or by its measure of value to the company. Asset value is usually done using a quantitative (monetary) value.

Authentication—1. The act of verifying identity, i.e., user, system

Scope notes: Risk: Can also refer to the verification of the correctness of a piece of data.

2. The act of verifying the identity of a user, the user's eligibility to access computerized information

Scope notes: Assurance: Authentication is designed to protect against fraudulent logon activity. It can also refer to the verification of the correctness of a piece of data.

Authenticity—Undisputed authorship

Availability—Ensuring timely and reliable access to and use of information

Awareness—Being acquainted with, mindful of, conscious of and well informed on a specific subject, which implies knowing and understanding a subject and acting accordingly

B

Balanced scorecard (BSC)—Developed by Robert S. Kaplan and David P. Norton as a coherent set of performance measures organized into four categories that includes traditional financial measures, but adds customer, internal business process, and learning and growth perspectives

Business case—Documentation of the rationale for making a business investment, used both to support a business decision on whether to proceed with the investment and as an operational tool to support management of the investment through its full economic life cycle

Business continuity—Preventing, mitigating and recovering from disruption

> **Scope notes:** The terms "business resumption planning," "disaster recovery planning" and "contingency planning" also may be used in this context; they focus on recovery aspects of continuity, and for that reason the 'resilience' aspect should also be taken into account.

Business continuity plan (BCP)—A plan used by an enterprise to respond to disruption of critical business processes. Depends on the contingency plan for restoration of critical systems

Business goal—The translation of the enterprise's mission from a statement of intention into performance targets and results

Business impact—The net effect, positive or negative, on the achievement of business objectives

Business impact analysis/assessment—Evaluating the criticality and sensitivity of information assets. An exercise that determines the impact of losing the support of any resource to an enterprise, establishes the escalation of that loss over time, identifies the minimum resources needed to recover, and prioritizes the recovery of processes and the supporting system.

> **Scope notes:** This process also includes addressing:
> - Income loss
> - Unexpected expense
> - Legal issues (regulatory compliance or contractual)
> - Interdependent processes
> - Loss of public reputation or public confidence

Business objective—A further development of the business goals into tactical targets and desired results and outcomes

Business process owner—The individual responsible for identifying process requirements, approving process design and managing process performance

> **Scope note:** Must be at an appropriately high level in the enterprise and have authority to commit resources to process-specific risk management activities

Business risk—A probable situation with uncertain frequency and magnitude of loss (or gain)

C

Capability—An aptitude, competency or resource that an enterprise may possess or require at an enterprise, business function or individual level that has the potential, or is required, to contribute to a business outcome and to create value

Capability Maturity Model (CMM)—1. Contains the essential elements of effective processes for one or more disciplines. It also describes an evolutionary improvement path from ad hoc, immature processes to disciplined, mature processes with improved quality and effectiveness

2. CMM for software, from the Software Engineering Institute (SEI), is a model used by many enterprises to identify best practices useful in helping them assess and increase the maturity of their software development processes

Scope notes: CMM ranks software development enterprises according to a hierarchy of five process maturity levels. Each level ranks the development environment according to its capability of producing quality software. A set of standards is associated with each of the five levels. The standards for level one describe the most immature or chaotic processes and the standards for level five describe the most mature or quality processes. A maturity model that indicates the degree of reliability or dependency the business can place on a process achieving the desired goals or objectives. A collection of instructions that an enterprise can follow to gain better control over its software development process

Change management—A holistic and proactive approach to managing the transition from a current to a desired organizational state, focusing specifically on the critical human or "soft" elements of change

Scope notes: Includes activities such as culture change (values, beliefs and attitudes), development of reward systems (measures and appropriate incentives), organizational design, stakeholder management, human resources (HR) policies and procedures, executive coaching, change leadership training, team building and communications planning and execution

Cloud computing—Convenient, on-demand network access to a shared pool of resources that can be rapidly provisioned and released with minimal management effort or service provider interaction

Compensating control—An internal control that reduces the risk of an existing or potential control weakness resulting in errors and omissions

Computer emergency response team (CERT)—A group of people integrated at the enterprise with clear lines of reporting and responsibilities for standby support in case of an information systems emergency

This group will act as an efficient corrective control, and should also act as a single point of contact for all incidents and issues related to information systems.

Confidentiality—Preserving authorized restrictions on access and disclosure, including means for protecting privacy and proprietary information

Configuration management—The control of changes to a set of configuration items over a system life cycle

Control—The means of managing risk, including policies, procedures, guidelines, practices or organizational structures, which can be of an administrative, technical, management, or legal nature

Control risk—The risk that a material error exists that would not be prevented or detected on a timely basis by the system of internal controls (See *Inherent risk*).

Control risk self-assessment—A method/process by which management and staff of all levels collectively identify and evaluate risk and controls with their business areas. This may be under the guidance of a facilitator such as an auditor or risk manager.

Copyright—Protection of writings, recordings or other ways of expressing an idea. The idea itself may be common, but they way it was expressed is unique, such as a song or book

Culture—A pattern of behaviors, beliefs, assumptions, attitudes and ways of doing things

Scope notes: COBIT 5 perspective

D

Data classification—The assignment of a level of sensitivity to data (or information) that results in the specification of controls for each level of classification. Levels of sensitivity of data are assigned according to predefined categories as data are created, amended, enhanced, stored or transmitted. The classification level is an indication of the value or importance of the data to the enterprise.

Data classification scheme—An enterprise scheme for classifying data by factors such as criticality, sensitivity and ownership

Data custodian—The individual(s) and department(s) responsible for the storage and safeguarding of computerized data

Data owner—The individual(s), normally a manager or director, who has responsibility for the integrity, accurate reporting and use of computerized data

Demilitarized zone (DMZ)—A screened (firewalled) network segment that acts as a buffer zone between a trusted and untrusted network

Scope notes: A DMZ is typically used to house systems such as web servers that must be accessible from both internal networks and the Internet

Detective control— Exists to detect and report when errors, omissions and unauthorized uses or entries occur

Disaster recovery plan (DRP)—A set of human, physical, technical and procedural resources to recover, within a defined time and cost, an activity interrupted by an emergency or disaster

E

Encryption—The process of taking an unencrypted message (plaintext), applying a mathematical function to it (encryption algorithm with a key) and producing an encrypted message (ciphertext)

Encryption algorithm—A mathematically based function or calculation that encrypts/decrypts data

Enterprise resource planning (ERP) system—A packaged business software system that allows an enterprise to automate and integrate the majority of its business processes, share common data and practices across the entire enterprise, and produce and access information in a real-time environment

Scope notes: Examples of ERP include SAP, Oracle Financials and J.D. Edwards.

Enterprise risk management (ERM)—The discipline by which an enterprise in any industry assesses, controls, exploits, finances and monitors risk from all sources for the purpose of increasing the enterprise's short- and long-term value to its stakeholders

Event—Something that happens at a specific place and/or time

Event type—For the purpose of IT risk management, one of three possible sorts of events: threat event, loss event and vulnerability event

Scope notes: Being able to consistently and effectively differentiate the different types of events that contribute to risk is a critical element in developing good risk-related metrics and well-informed decisions. Unless these categorical differences are recognized and applied, any resulting metrics lose meaning and, as a result, decisions based on those metrics are far more likely to be flawed.

Evidence—1. Information that proves or disproves a stated issue

2. Information an auditor gathers in the course of performing an IS audit; relevant if it pertains to the audit objectives and has a logical relationship to the findings and conclusions it is used to support

Scope note: Audit perspective

F

Fallback procedures—A plan of action or set of procedures to be performed if a system implementation, upgrade or modification does not work as intended

Scope note: May involve restoring the system to its state prior to the implementation or change. Fallback procedures are needed to ensure that normal business processes continue in the event of failure and should always be considered in system migration or implementation

Feasibility study—A phase of a system development life cycle (SDLC) methodology that researches the feasibility and adequacy of resources for the development or acquisition of a system solution to a user need

Framework—A generally accepted, business-process-oriented structure that establishes a common language and enables repeatable business processes

Scope note: This term may be defined differently in different disciplines. This definition suits the purposes of this manual.

Frequency—A measure of the rate by which events occur over a certain period of time

G

Governance—Ensures that stakeholder needs, conditions and options are evaluated to determine balanced, agreed-on enterprise objectives to be achieved; setting direction through prioritization and decision making; and monitoring performance and compliance against agreed-on direction and objectives

Scope note: Conditions can include the cost of capital, foreign exchange rates, etc. Options can include shifting manufacturing to other locations, subcontracting portions of the enterprise to third parties, selecting a product mix from many available choices, etc.

Governance enabler—Something (tangible or intangible) that assists in the realization of effective governance

Scope notes: COBIT 5 perspective

Governance of enterprise IT—A governance view that ensures that information and related technology support and enable the enterprise strategy and the achievement of enterprise objectives; this also includes the functional governance of IT, i.e., ensuring that IT capabilities are provided efficiently and effectively.

Scope notes: COBIT 5 perspective

I

Impact—Magnitude of loss resulting from a threat exploiting a vulnerability

Impact analysis—A study to prioritize the criticality of information resources for the enterprise based on costs (or consequences) of adverse events

In an impact analysis, threats to assets are identified and potential business losses determined for different time periods. This assessment is used to justify the extent of safeguards that are required and recovery time frames. This analysis is the basis for establishing the recovery strategy.

Impact assessment—A review of the possible consequences of a risk

Scope notes: See also *Impact analysis*

Incident—Any event that is not part of the standard operation of a service and that causes, or may cause, an interruption to, or a reduction in, the quality of that service

Information security— Ensures that within the enterprise, information is protected against disclosure to unauthorized users (confidentiality), improper modification (integrity), and non-access when required (availability)

Information systems (IS)—The combination of strategic, managerial and operational activities involved in the gathering, processing, storing, distributing and use of information, and its related technologies

Scope notes: Information systems are distinct from information technology (IT) in that an information system has an IT component that interacts with the process components

Information technology (IT)—The hardware, software, communication and other facilities used to input, store, process, transmit and output data in whatever form

Infrastructure as a Service (IaaS)—Offers the capability to provision processing, storage, networks and other fundamental computing resources, enabling the customer to deploy and run arbitrary software, which can include operating systems (OSs) and applications

Inherent risk—The risk level or exposure without taking into account the actions that management has taken or might take (e.g., implementing controls)

Integrity— The guarding against improper information modification or destruction, and includes ensuring information nonrepudiation and authenticity

Intellectual property—Intangible assets that belong to an enterprise for its exclusive use

Internal controls—The policies, procedures, practices and organizational structures designed to provide reasonable assurance that the business objectives will be achieved and undesired events will be prevented or detected and corrected

IT architecture— Description of the fundamental underlying design of the IT components of the business, the relationships among them, and the manner in which they support the enterprise's objectives

IT infrastructure—The set of hardware, software and facilities that integrates an enterprise's IT assets

Scope note: Specifically, the equipment (including servers, routers, switches, and cabling), software, services and products used in storing, processing, transmitting and displaying all forms of information for the organization's users

IT-related incident—An IT-related event that causes an operational, developmental and/or strategic business impact

IT risk—The business risk associated with the use, ownership, operation, involvement, influence and adoption of IT within an enterprise

IT risk issue—1. An instance of an IT risk

2. A combination of control, value and threat conditions that impose a noteworthy level of IT risk

IT risk profile—A description of the overall (identified) IT risk to which the enterprise is exposed

IT risk register—A repository of the key attributes of potential and known IT risk issues. Attributes may include name, description, owner, expected/actual frequency, potential/actual magnitude, potential/actual business impact and disposition

IT risk scenario—The description of an IT-related event that can lead to a business impact

IT strategic plan—A long-term plan (i.e., three- to five-year horizon) in which business and IT management cooperatively describe how IT resources will contribute to the enterprise's strategic objectives (goals)

IT tactical plan—A medium-term plan (i.e., six- to 18-month horizon) that translates the IT strategic plan direction into required initiatives, resource requirements and ways in which resources and benefits will be monitored and managed

K

Key performance indicator (KPI)—A measure that determines how well the process is performing in enabling the goal to be reached

Scope Note: A lead indicator of whether a goal will likely be reached, and a good indicator of capabilities, practices and skills. It measures an activity goal, which is an action that the process owner must take to achieve effective process performance

Key risk indicator (KRI)—A subset of risk indicators that are highly relevant and possess a high probability of predicting or indicating important risk

Scope note: See *Risk indicator*

L

Lag indicator—Metrics for achievement of goals-An indicator relating to the outcome or result of an enabler

Scope note: This indicator is only available after the facts or events.

Lead indicator—Metrics for application of good practice-An indicator relating to the functioning of an enabler

Scope note: This indicator will provide an indication on possible outcome of the enabler

Likelihood—The probability of something happening

Loss event— Any event where a threat event results in loss

Scope note: From Jones, J.; "FAIR Taxonomy," Risk Management Insight, USA, 2008

M

Magnitude— A measure of the potential severity of loss or the potential gain from realized events/scenarios

Management—Plans, builds, runs and monitors activities in alignment with the direction set by the governance body to achieve the enterprise objectives

N

Nondisclosure agreement (NDA)— A legal contract between at least two parties that outlines confidential materials that the parties wish to share with one another for certain purposes, but wish to restrict from generalized use; a contract through which the parties agree not to disclose information covered by the agreement

Scope notes: Also called a confidential disclosure agreement (CDA), confidentiality agreement or secrecy agreement. An NDA creates a confidential relationship between the parties to protect any type of trade secret. As such, an NDA can protect non-public business information. In the case of certain governmental entities, the confidentiality of information other than trade secrets may be subject to applicable statutory requirements, and in some cases may be required to be revealed to an outside party requesting the information. Generally, the governmental entity will include a provision in the contract to allow the seller to review a request for information that the seller identifies as confidential and the seller may appeal such a decision requiring disclosure. NDAs are commonly signed when two companies or individuals are considering doing business together and need to understand the processes used in one another's businesses solely for the purpose of evaluating the potential business relationship. NDAs can be "mutual," meaning that both parties are restricted in their use of the materials provided, or they can only restrict a single party. It is also possible for an employee to sign an NDA or NDA-like agreement with a company at the time of hiring; in fact, some employment agreements will include a clause restricting "confidential information" in general.

Nonrepudiation—The assurance that a party cannot later deny originating data; provision of proof of the integrity and origin of the data and that can be verified by a third party

Scope notes: A digital signature can provide nonrepudiation

O

Objectivity—The ability to exercise judgment, express opinions and present recommendations with impartiality

Operational level agreement (OLA)—An internal agreement covering the delivery of services that support the IT organization in its delivery of services

Owner— Individual or group that holds or possesses the rights of and the responsibilities for an enterprise, entity or asset

Scope notes: Examples: process owner, system owner

P

Patent—Protection of research and ideas that led to the development of a new, unique and useful product to prevent the unauthorized duplication of the patented item

Penetration testing—A live test of the effectiveness of security defenses through mimicking the actions of real-life attackers

Performance indicators—A set of metrics designed to measure the extent to which performance objectives are being achieved on an ongoing basis

Scope note: Performance indicators can include service level agreements (SLA), critical success factors (CSF), customer satisfaction ratings, internal or external benchmarks, industry best practices and international standards

Platform as a Service (PaaS)—Offers the capability to deploy onto the cloud infrastructure customer-created or -acquired applications that are created using programming languages and tools supported by the provider

Policy—1. Generally, a document that records a high-level principle or course of action that has been decided on

The intended purpose is to influence and guide both present and future decision making to be in line with the philosophy, objectives and strategic plans established by the enterprise's management teams.

Scope notes: In addition to policy content, policies need to describe the consequences of failing to comply with the policy, the means for handling exceptions, and the manner in which compliance with the policy will be checked and measured.

2. Overall intention and direction as formally expressed by management

Scope notes: COBIT 5 perspective

Portfolio—A grouping of "objects of interest" (investment programs, IT services, IT projects, other IT assets or resources) managed and monitored to optimize business value. (The investment portfolio is of primary interest to Val IT. The IT service, project, asset and other resource portfolios are of primary interest to COBIT.)

Preventive control—An internal control that is used to avoid undesirable events, errors and other occurrences that an enterprise has determined could have a negative material effect on a process or end product

Privilege—The level of trust with which a system object is imbued

Problem—In IT, the unknown underlying cause of one or more incidents

Problem escalation procedure—The process of escalating a problem up from junior to senior support staff, and ultimately to higher levels of management

Scope notes: Problem escalation procedure is often used in help desk management, when an unresolved problem is escalated up the chain of command, until it is solved.

Program—A structured grouping of interdependent projects that is both necessary and sufficient to achieve a desired business outcome and create value. These projects could include, but are not limited to, changes in the nature of the business, business processes and the work performed by people as well as the competencies required to carry out the work, the enabling technology and the organizational structure.

Project—A structured set of activities concerned with delivering a defined capability (that is necessary, but not sufficient, to achieve a required business outcome) to the enterprise, based on an agreed-on schedule and budget

Project portfolio—The set of projects owned by a company

Scope note: It usually includes the main guidelines relative to each project, including objectives, costs, time lines and other information specific to the project

Q

Qualitative risk analysis—Defines risk using a scale or comparative values (i.e., defining risk factors in terms of high/medium/low or on a numeric scale from 1 to 10). It is based on judgment, intuition and experience rather than on financial values

Quantitative risk analysis—The use of numerical and statistical techniques to calculate likelihood and impact of risk. It uses financial data, percentages and ratios to provide an approximate measure of the magnitude of impact in financial terms

R

RACI chart—Illustrates who is Responsible, Accountable, Consulted and Informed within an organizational framework

Recovery point objective (RPO)—Determined based on the acceptable data loss in case of a disruption of operations. It indicates the earliest point in time to which it is acceptable to recover the data. The RPO effectively quantifies the permissible amount of data loss in case of interruption.

Recovery strategy—An approach by an enterprise that will ensure its recovery and continuity in the face of a disaster or other major outage

> **Scope notes:** Plans and methodologies are determined by the enterprise's strategy. There may be more than one methodology or solution for an enterprise's strategy. Examples of methodologies and solutions include contracting for hot site or cold site, building an internal hot site or cold site, identifying an alternate work area, a consortium or reciprocal agreement, contracting for mobile recovery or crate and ship, and many others.

Recovery testing—A test to check the system's ability to recover after a software or hardware failure

Recovery time objective (RTO)—The amount of time allowed for the recovery of a business function or resource after a
disaster occurs

Residual risk—The remaining risk after management has implemented risk response

Resilience—The ability of a system or network to resist failure or to recover quickly from any disruption, usually with minimal recognizable effect

Return on investment (ROI)—A measure of operating performance and efficiency, computed in its simplest form by dividing net income by the total investment over the period being considered

Risk—The combination of the probability of an event and its consequence. (ISO/IEC73)

Risk acceptance—If the risk is within the enterprise's risk tolerance or if the cost of otherwise mitigating the risk is higher than the potential loss, the enterprise can assume the risk and absorb any losses

Risk aggregation—The process of integrating risk assessments at a corporate level to obtain a complete view on the overall risk for the enterprise

Risk analysis—1. A process by which frequency and magnitude of IT risk scenarios are estimated.

2. The initial steps of risk management: analyzing the value of assets to the business, identifying threats to those assets and evaluating how vulnerable each asset is to those threats

> **Scope notes:** It often involves an evaluation of the probable frequency of a particular event, as well as the probable impact of that event

Risk appetite—The amount of risk, on a broad level, that an entity is willing to accept in pursuit of its mission

Risk assessment—A process used to identify and evaluate risk and its potential effects

> **Scope note:** Includes assessing the critical functions necessary for an enterprise to continue business operations, defining the controls in place to reduce organization exposure and evaluating the cost for such controls. Risk analysis often involves an evaluation of the probabilities of a particular event.

Risk avoidance—The process for systematically avoiding risk, constituting one approach to managing risk

Risk culture—The set of shared values and beliefs that governs attitudes toward risk-taking, care and integrity, and determines how openly risk and losses are reported and discussed

Risk evaluation—The process of comparing the estimated risk against given risk criteria to determine the significance of the risk (ISO/IEC Guide 73:2002)

Risk factor—A condition that can influence the frequency and/or magnitude and, ultimately, the business impact of IT-related events/scenarios

Risk identification—The process of determining and documenting the risk that an enterprise faces. The identification of risk is based on the recognition of threats, vulnerabilities, assets and controls in the enterprise's operational environment

Risk impact—The calculation of the amount of loss or damage that an organization may incur due to a risk event

Risk indicator—A metric capable of showing that the enterprise is subject to, or has a high probability of being subject to, a risk that exceeds the defined risk appetite

Risk management—1. The coordinated activities to direct and control an enterprise with regard to risk

Scope notes: In the International Standard, the term "control" is used as a synonym for "measure" (ISO/IEC Guide 73:2002)

2. One of the governance objectives. Entails recognizing risk; assessing the impact and likelihood of that risk; and developing strategies, such as avoiding the risk, reducing the negative effect of the risk and/or transferring the risk, to manage it within the context of the enterprise's risk appetite

Scope notes: COBIT 5 perspective

Risk map—A (graphic) tool for ranking and displaying risk by defined ranges for frequency and magnitude

Risk mitigation—The management of risk through the use of countermeasures and controls

Risk portfolio view—1. A method to identify interdependencies and interconnections among risk, as well as the effect of risk responses on multiple types of risk

2. A method to estimate the aggregate impact of multiple types of risk (e.g., cascading and coincidental threat types/scenarios, risk concentration/correlation across silos) and the potential effect of risk response across multiple types of risk

Risk owner—The person in whom the organization has invested the authority and accountability for making risk-based decisions and who owns the loss associated with a realized risk scenario

Scope notes: The risk owner may not be responsible for the implementation of risk treatment.

Risk scenario—The tangible and assessable representation of risk

Scope notes: One of the key information items needed to identify, analyze and respond to risk (COBIT 5 Process APO12)

Risk statement—A description of the current conditions that may lead to the loss; and a description of the loss Source: Software Engineering Institute (SEI)

Scope notes: For a risk to be understandable, it must be expressed clearly. Such a treatment must include a description of the current conditions that may lead to the loss; and a description of the loss.

Risk tolerance—The acceptable level of variation that management is willing to allow for any particular risk as the enterprise pursues its objectives

Risk transfer—The process of assigning risk to another enterprise, usually through the purchase of an insurance policy or by outsourcing the service

Scope notes: Also known as risk sharing

Root cause analysis—A process of diagnosis to establish the origins of events, which can be used for learning from consequences, typically from errors and problems

Scope creep—Also called requirement creep, this refers to uncontrolled changes in a project's scope

Scope notes: Scope creep can occur when the scope of a project is not properly defined, documented and controlled. Typically, the scope increase consists of either new products or new features of already approved products. Hence, the project team drifts away from its original purpose. Because of one's tendency to focus on only one dimension of a project, scope creep can also result in a project team overrunning its original budget and schedule. For example, scope creep can be a result of poor change control, lack of proper identification of what products and features are required to bring about the achievement of project objectives in the first place, or a weak project manager or executive sponsor

Segregation/separation of duties (SoD)—A basic internal control that prevents or detects errors and irregularities by assigning to separate individuals the responsibility for initiating and recording transactions and for the custody of assets

Scope notes: Segregation/separation of duties is commonly used in large IT organizations so that no single person is in a position to introduce fraudulent or malicious code without detection

Service level agreement (SLA)—An agreement, preferably documented, between a service provider and the customer(s)/user(s) that defines minimum performance targets for a service and how they will be measured

Slack time (float)—Time in the project schedule, the use of which does not affect the project's critical path; the minimum time to complete the project based on the estimated time for each project segment and their relationships

Scope note: Slack time is commonly referred to as "float" and generally is not "owned" by either party to the transaction

Software as a Service (SaaS)—Offers the capability to use the provider's applications running on cloud infrastructure. The applications are accessible from various client devices through a thin client interface such as a web browser (e.g., web-based email)

Standard—A mandatory requirement, code of practice or specification approved by a recognized external standards organization, such as the International Organization for Standardization (ISO)

Statement of work (SOW)—A formal document that captures and defines the work activities, deliverables, and time line a vendor must execute in performance of specified work for a client

The SOW usually includes detailed requirements and pricing, with standard regulatory and governance terms and conditions

Strategic planning—The process of deciding on the enterprise's objectives, on changes in these objectives, and the policies to govern their acquisition and use

System development life cycle (SDLC)—The phases deployed in the development or acquisition of a software system

> **Scope notes:** SDLC is an approach used to plan, design, develop, test and implement an application system or a major modification to an application system. Typical phases of SDLC include the feasibility study, requirements study, requirements definition, detailed design, programming, testing, installation and post-implementation review, but not the service delivery or benefits realization activities.

T

Threat—Anything (e.g., object, substance, human) that is capable of acting against an asset in a manner that can result in harm

> **Scope note:** A potential cause of an unwanted incident (ISO/IEC 13335)

Threat agent—Methods and things used to exploit a vulnerability

> **Scope notes:** Examples include determination, capability, motive and resources

Threat analysis—An evaluation of the type, scope and nature of events or actions that can result in adverse consequences; identification of the threats that exist against enterprise assets

> **Scope notes:** The threat analysis usually defines the level of threat and the likelihood of it materializing

Threat event—Any event where a threat element/actor acts against an asset in a manner that has the potential to directly result in harm

Threat vector—The path or route used by the adversary to gain access to the target

Trademark—A sound, color, logo, saying or other distinctive symbol that is closely associated with a certain product or company

V

Vulnerability—A weakness in the design, implementation, operation or internal control of a process that could expose the system to adverse threats from threat events

Vulnerability analysis—A process of identifying and classifying vulnerabilities

Vulnerability event—Any event where a material increase in vulnerability results. Note that this increase in vulnerability can result from changes in control conditions or from changes in threat capability/force

> **Scope notes:** From Jones, J.; "FAIR Taxonomy," Risk Management Insight, USA, 2008

Vulnerability scanning—An automated process to proactively identify security weaknesses in a network or individual system

Page intentionally left blank

Index

A slash (/) indicates that the terms are synonymous within this manual.

"See also" indicates that the terms are related or relevant to one another.

A

B

C

D

E

F

G

H

I

CRISC
Certified in Risk and Information Systems Control
An ISACA Certification

S

T

U

V

W

Page intentionally left blank

EVALUATION

ISACA continuously monitors the swift and profound professional, technological and environmental advances affecting risk and IS control professionals. Recognizing these rapid advances, the *CRISC™ Review Manual* is updated annually.

To assist ISACA in keeping abreast of these advances, please take a moment to evaluate the *CRISC™ Review Manual 6th Edition*. Such feedback is valuable to fully serve the profession and future CRISC exam registrants.

To complete the evaluation on the web site, please go to *www.isaca.org/studyaidsevaluation*.

Thank you for your support and assistance.